T0336517

Nuclear Data

A collective motion view

Online at: https://doi.org/10.1088/978-0-7503-5643-5

IOP Series in Nuclear Spectroscopy and Nuclear Structure

Series Editors

John Wood, Georgia Institute of Technology, USA
Kristiaan Heyde, Ghent University, Belgium
David Jenkins, University of York, UK

About the Series:

The IOP Series in Nuclear Spectroscopy and Nuclear Structure provides up-to-date coverage of theoretical and experimental topics in all areas of modern nuclear spectroscopy and structure. Books in the series range from student primers, graduate textbooks, research monographs and practical guides to meet the needs of students and scientists.

All aspects of nuclear spectroscopy and structure research are included, for example:
- Nuclear systematics and data
- Ab-initio models
- Shell model-based descriptions
- Nuclear collective models
- Nuclear symmetries and algebraic descriptions
- Many-body aspects of nuclear structure
- Quantum mechanics for nuclear structure study
- Spectroscopy with gamma-rays, charged particles and neutrons following radioactive decay and reactions
- Spectroscopy of rare isotopes
- Detectors employed for gamma-ray, neutron and charged-particle detection in nuclear spectroscopy and related societal applications.

A full list of titles published in this series can be found here: https://iopscience.iop.org/bookListInfo/iop-series-in-nuclear-spectroscopy-and-nuclear-structure#series.

Nuclear Data

A collective motion view

David Jenkins

School of Physics, Engineering and Technology, University of York, York YO10 5DD

John L Wood

School of Physics, Georgia Institute of Technology, Atlanta, Georgia 30332, USA

IOP Publishing, Bristol, UK

Multimedia content is available for this book from https://doi.org/10.1088/978-0-7503-5643-5.

ISBN   978-0-7503-5643-5 (ebook)
ISBN   978-0-7503-5641-1 (print)
ISBN   978-0-7503-5644-2 (myPrint)
ISBN   978-0-7503-5642-8 (mobi)

DOI   10.1088/978-0-7503-5643-5

Version: 20231201

IOP ebooks

British Library Cataloguing-in-Publication Data: A catalogue record for this book is available from the British Library.

Published by IOP Publishing, wholly owned by The Institute of Physics, London

IOP Publishing, No.2 The Distillery, Glassfields, Avon Street, Bristol, BS2 0GR, UK

US Office: IOP Publishing, Inc., 190 North Independence Mall West, Suite 601, Philadelphia, PA 19106, USA

Contents

Preface

Nuclear data show many different faces, and our task is to provide a means to navigate this data base. In *Nuclear Data: A primer* (https://iopscience.iop.org/book/mono/978-0-7503-2674-2) we aimed to provide an orientation to assist the newcomer; in this and the following volumes in the series we aim to introduce the reader to the nuclear structure behind the data. The data base undergoes significant changes on a decadal scale; ideas for interpretation and exploration can change on a yearly basis.

We divide the material presented into a minimal set of structural classes. We place collective motion at the beginning because it dominates nuclear structure. While the shell model is at the foundational level of nuclear structure, its independent-particle degrees of freedom rarely dominate a given nucleus. When they appear to do so, they are still 'dressed' in so-called correlations. At the meeting point between collective degrees of freedom and independent-particle degrees of freedom one encounters a major research frontier in the study of the nuclear many-body problem.

In this volume we endeavour to address both well-classified features of nuclear data and aspects of nuclear data that lack a consensus regarding what is going on. In this way, from our perspective, a most important message is: the saga of nuclear structure exploration is moving from 'the end of its beginning to the beginning of its in-depth *systematic* investigation'.

By presenting the chapters as questions, we aim to achieve two outcomes: (a) none of the structural concepts are thoroughly elucidated and so the reader should be encouraged to be skeptical of interpretations that we make; (b) it is critical for the reader to adopt the view that there is much work to be done.

Nuclear collectivity can be subdivided into rotations and vibrations. Even this division has been the subject of debate over the decades: this is due, in part, to inadequate data; but it has resulted in rotational–vibrational models that blur the distinction. We will adhere to a data-based view and only present the simplest rotational and vibrational models for comparison. However, even basic model concepts such as a moment of inertia are questionable when data are considered in detail. Further, it will be found that the existence of vibrations at low energy in nuclei remain elusive. Entering the classification of nuclear collectivity is the manifestation of coexisting shapes in nuclei. This category includes nuclear structures with the largest deformations known, albeit not in association with nuclear ground states.

Shape coexistence deals with the issue of structures with different shapes occurring in a single nuclear species, (A, Z, N). The related issue of (coexisting) cluster structure in nuclei remains poorly defined. The focus on coexistence of structures in nuclei has slowly expanded over the past forty years; but is still barely out of its 'infancy'. Thus, little is known about the extent of its occurrence: can shape coexistence occur in all nuclei? Just how many different shapes can a given nucleus have?

The term 'collective' implies a coherent or cooperative behaviour of the constituents of a many-body system. In classical mechanics this is formalized under the title 'rigid-body' motion. The most dramatic example is the rotation of an extended rigid body. But there are also collective motions in fluids: it is remarkable that pictures of the atmosphere of the planet Jupiter exhibit patterns recognizable as 'turbulent' fluid flow that appear identical to that observed when you stir cream into a cup of black coffee, a scale difference of $\sim 10^9$. Nuclei exhibit properties that are both rigid-body-like and fluid-like in their collective behaviour, and we will find that much remains to be learned; but some simple quantum mechanical models involving rotations work amazingly well, albeit with parameters that if interpreted as moments of inertia bear no resemblance to anything observed classically under this name.

Classically, an extended rigid body can exhibit rotations and vibrations—these are manifestly collective modes that involve the entire body. (We do not discuss 'infinite' systems here as there are few useful analogies with nuclei.) Rotations are easily visualized, literally, although the free-space rotation of a rigid asymmetrical object can exhibit some very peculiar behaviour[1]. Vibrations often cannot easily be seen in extended solid bodies, but sometimes can be heard! In the details, one talks of the so-called 'normal modes' of vibration of a given extended body. Vibrations of fluids are visible for surface modes but not directly for bulk modes. A useful 'bridge' into the quantum mechanical world is molecules, which exhibit quantized rotations and vibrations with extraordinarily rich structures. This body of mechanical knowledge has naturally found its way into trying to understand collective motion in nuclei: sometimes the analogies have proven useful; sometimes they have seriously obstructed the path to understanding collective nuclear motion. The focus chosen on the details of nuclear data presented herein is aimed at exploring what can and cannot be said about nuclear rotations and vibrations.

References

[1] Peterson C and Schwalm W 2021 Euler's rigid rotators, Jacobi elliptic functions and the Dzhanibekov or tennis racket effect *Am. J. Phys.* **89** 349

[2] Mardešić P *et al* 2020 Geometric origin of the tennis racket effect *Phys. Rev. Lett.* **125** 064301

[3] Van Damme L *et al* 2017 Linking the rotation of a rigid body to the Schrödinger equation: the quantum tennis racket effect and beyond *Sci. Rep.* **7** 3998

[4] Van Damme L, Mardešić P and Sugny D 2017 The tennis racket effect in a three-dimensional rigid body *Physica* D **338** 17

[5] Ashbaugh M S, Chicone C C and Cushman R H 1991 The twisting tennis racket *J. Dyn. Differ. Equ.* **3** 67

[6] Harter W G and Kim C C 1976 Singular motions of asymmetric rotators *Am. J. Phys.* **44** 1080

---

[1] Under the title of the 'tennis racket effect' (also called the 'Dzhanibekov effect', after the Russian cosmonaut, Vladimir Dzhanibekov who beautifully demonstrated the effect in space using a spinning wing nut, which intermittently makes 180° flips of its rotation axis) one observes the effect when holding a tennis racket 'flat' and tossing the handle upwards and away from you. Note, the idea is to catch the object after it has completed one or more rotations—if you are not good at this, do not try it with a book, or a smart phone if standing on a concrete floor. The physics behind this is remarkable, see, e.g., [1–6]. Further note, if a supposed rigid axially symmetric rotor is not quite axially rigid, the effect will appear, much to the astonishment and cost of one satellite launch team. (The satellite was Explorer One and was 'pencil shaped'. It was initially set spinning about its minor axis ['pencil lead' axis], but due to flexible 'whisker' antennae it soon changed to rotation about its major axis. This essentially ended its ability to transmit data and receive commands.)

Author biographies

David Jenkins

David Jenkins is Head of the Nuclear Physics Group at the University of York, UK. He is also a Fellow of the Institute of Advanced Study, University of Strasbourg (USIAS) and an Extraordinary Professor of the University of Western Cape in South Africa. His research in experimental nuclear physics focusses on several topics such as nuclear astrophysics, clustering in nuclei and the study of proton-rich nuclei. In recent years, he has developed a strong strand of applications-related research with extensive industrial collaboration. He has led the development of bespoke radiation detectors for homeland security, nuclear decommissioning, borehole logging and medical applications.

John Wood

John Wood is a Professor Emeritus in the School of Physics at Georgia Institute of Technology. He continues to collaborate on research projects in both experimental and theoretical nuclear physics. Special research interests include nuclear shapes and systematics of nuclear structure.

IOP Publishing

Nuclear Data
A collective motion view
David Jenkins and John L Wood

Chapter 1

How well defined are rotations in nuclei?

Rotational states in nuclei, their symmetries and their quantum numbers are introduced. The symmetric top model and its energies and electric quadrupole, E2 properties are confronted with data. The roles of spherical tensor operators, Clebsch–Gordan coefficients and the Wigner–Eckart theorem are explained. The subtle nature of the quantum mechanical uncertainty associated with rotations is illustrated. Rotational-particle coupling is defined, its main features are explored, and a data-based view is presented. A first look is taken at the peculiarities of nuclear moments of inertia.

Concepts: energy patterns, total spin quantum number, K quantum number, state vectors, electric quadrupole ($E2$) properties, symmetric top model, intrinsic quadrupole (Q_0) parameter, spherical tensor operators, Clebsch–Gordan coefficients, Wigner–Eckart theorem, body frame, laboratory frame, quantum mechanical uncertainty, $E2$ transitions, lifetimes, rotational-particle coupling, Coriolis and recoil terms, rotation alignment, moment of inertia, deformation parameter, super-deformed band, rigid and irrotational flow.

Learning outcomes: the key data view from this chapter is the widespread evidence for simple symmetric rotor behaviour in many nuclei. Liquid-drop behaviour of nuclei is contradicted by the emergence of constant intrinsic quadrupole moments with increasing nuclear rotational angular momentum; indeed, constancy of intrinsic quadrupole moments may be realized at the few percent level. Such constancy is not matched by rotational energy patterns, i.e. nuclei do not exhibit simple analogues of classical moments of inertia. Furthermore, coupling an odd particle spin to a rotor core angular momentum does not show the quantum mechanical analogue of classical Coriolis effects.

Rotations are widely identified in nuclei and their existence is not in question. But just what is rotating is an open question: beyond a conformity to the simplest model expressions, data indicate a variety of features for which there are only emerging

doi:10.1088/978-0-7503-5643-5ch1

ideas and there is no consensus on interpretation[1]. In this chapter, we progressively look at data to see what we can and cannot say about nuclear rotation.

1.1 Even–even nuclei: energies and electric quadrupole, $E2$ properties

The simplest criterion for assessing nuclear rotation is a comparison of the energies of ground-state bands of even–even nuclei with the formula,

$$E_I = AI(I + 1), \tag{1.1}$$

where I is the spin of the state, $I = 0, 2, 4, 6, \ldots$ and A is a free parameter. This leads to the familiar ratio test, $E_4/E_2 = 3.333$ (also $E_6/E_2 = 7.000$, etc). Excited bands and bands in odd and odd–odd nuclei can be tested in a similar manner, allowing for a band-head excitation and non-zero spin of the band head. At a deeper level, a band can be assigned a K quantum number and electromagnetic properties can be assigned to states in a band, e.g., as an intrinsic quadrupole moment, Q_0.

The simplicity[2] of the energy patterns defined above for the ground-state bands of even–even nuclei demands a simple explanation. It is manifested in the symmetric top model[3] which is depicted in figure 1.1. This view of deformed even–even nuclei steps beyond all the details of the tens to hundreds of nucleons involved and recognizes that there is an axis of rotational symmetry and a plane of reflection symmetry at right angles to this axis. The axis of rotational symmetry has the consequence that the projection of the total nuclear spin on this axis is a constant of motion. This can be assigned a quantum number, traditionally labelled by K. The plane of reflection symmetry dictates that there is a two-fold energy degeneracy with respect to the K quantum number: the states $|+K\rangle$ and $|-K\rangle$ are indistinguishable energetically. Fundamentally, when such a degeneracy arises in a quantum mechanical system, a linear combination must be adopted to describe the system. Failure to do this would result in writing a state vector, e.g. '$|+K\rangle$' that would

---

[1] Caution is needed when attempting to conceptualize quantized rotations: quantization results in stationary states, often visualized as standing waves. In atoms, this results in the iconic atomic electron density distributions, labelled by angular momentum quantum numbers, represented using spherical harmonics. Recent studies of ultrafast pulsed laser imaging of molecules report on observations that endeavour to sharpen our view of quantized rotations [1–3]; but it is impossible to escape the limitation of only ever forming probability density 'images'. Nevertheless, we will revisit this issue in various ways in the coming narrative.

[2] The simplicity of the 'even-spins only' property of symmetric tops with a plane of reflection symmetry is worth a deeper consideration. These are quantum mechanical systems, ranging from molecules to nuclei, possessing a fundamental symmetry. In the spectroscopy of such systems, the odd-spin states are 'missing'. This has been the basis of searches for exotica, i.e. physical entities (particles) that have found a 'refuge' in symmetric top systems with the result that odd-spin states are manifested (weakly, the species are presumably rare) in the spectra. Such systems also provide a 'laboratory' for testing the symmetrization postulate applied to identical particles: just suppose that so-called 'identical' particles are not quite identical. We recommend to the reader the following: [4–8].

[3] The symmetric top model is a model of a rigid rotor with two components of the inertia tensor equal. This is characteristic of extended rigid bodies with an axis of rotational symmetry. A symmetric top can have a plane of reflection symmetry or reflection asymmetry through the centre of mass of the body at right angles to the axis of rotational symmetry.

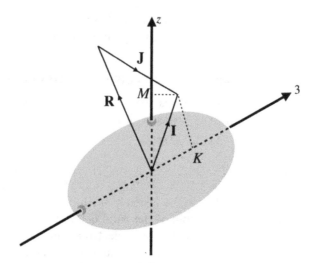

Figure 1.1. The quantum numbers of the symmetric top model superimposed on a schematic view of an axially symmetric deformed object, with a plane of reflection symmetry through the centre of mass at right angles to the symmetry axis. The symmetry axis is in the body-fixed frame and is labelled '3'. A laboratory-fixed frame axis, labelled z is also shown. Quantization of the total spin of the system, I can only be assigned sharp values for one directional component, independently with respect to the body-fixed axis and with respect to the laboratory-fixed axis, i.e. the 3-axis and the z-axis are not in any fixed relationship to each other. This relationship is not defined for a quantum mechanical rotor: a depiction of this uncertainty is attempted in figure 1.6. Thus, there are quantum numbers K (body frame) and M (lab frame), which independently range over the values $+I, +I-1, \ldots, -I$, respectively. The total spin $\mathbf{I}$ of the system has an intrinsic contribution $\mathbf{J}$ and a rotational contribution $\mathbf{R}$; these are discussed further in the text. The vector $\mathbf{R}$ is directed at right angles to the 3-axis.

imply the direction of K was known to be 'positive', which is an impossibility: there is no measurement which can establish this. Thus, we write

$$|IMK\rangle = \frac{\{|IM, +K\rangle + (-1)^{I+K}|IM, -K\rangle\}}{\sqrt{2}}, \qquad (1.2)$$

where I is the total nuclear spin, K is the projection of I on the nuclear symmetry axis, and M is a directional component of spin in the laboratory-fixed frame of reference. The phase factor appearing in equation (1.2) applies to a plane of reflection symmetry; if the body has a plane of reflection asymmetry (through the centre of mass at right angles to the symmetry axis), there is a minus sign instead of a plus sign between the two state kets. The distinction of a body-fixed frame from the laboratory-fixed frame is discussed in detail in section 1.2. The key point here is that if $K = 0$, the state vector components $|IM, +K\rangle$ and $|IM, -K\rangle$ are indistinguishable; thus, in equation (1.2) only even values of I can occur.

Energies are observed to deviate from equation (1.1), in a smooth systematic manner. There are many formulae that attempt to describe these deviations. The simplest is

$$E_I = AI(I+1) + BI^2(I+1)^2, \qquad (1.3)$$

where A and B are free parameters. There are other formulae expressed in terms of polynomials in I, formulae expressed using what is termed a rotational frequency, and formulae invoking parameterizations of the moment of inertia (manifested in the A parameter, above); these formulae generally have no basis in quantum mechanics, i.e., they are motivated from classical mechanics, or they are even purely mathematical with no physics motivation. To enumerate this aspect of the modelling of nuclear rotations would be laborious, inconclusive with respect to the physics involved, and indeed a distraction from the focus that we adopt herein. None of these phenomenological refinements describe the observed energies exactly. We will return to energy patterns in rotational bands shortly.

Electromagnetic properties show a remarkable conformity to the simplest quantum mechanical models in nearly all nuclei which exhibit collective behaviour. This is the starting point for the focus that we adopt. The prototype relationships are for $B(E2)$ values of transitions and electric quadrupole moments in the ground-state bands of strongly deformed even–even nuclei, expressed as matrix elements of the electric quadrupole operator, $T(E2)$.

For $B(E2)$ values, in ground-state bands of even–even nuclei, which have $K = 0$,

$$B(E2; I \rightarrow I - 2) = \frac{\langle I0|T(E2)|I - 2, 0\rangle^2}{(2I + 1)}, \tag{1.4}$$

where

$$\langle I0|T(E2)|I - 2, 0\rangle = (2I + 1)^{1/2}(5/16\pi)^{1/2}\langle I020|I - 2, 0\rangle eQ_0, \tag{1.5}$$

$\langle I020|I - 2, 0\rangle$ is a Clebsch–Gordan coefficient, e is the fundamental unit of electric charge and Q_0 is a model parameter describing the intrinsic quadrupole moment of the nucleus. Note, cf equation (1.2), the states are expressed in terms of the I and K quantum numbers, viz. $|IK\rangle$, with the M quantum number omitted because these processes are independent of the orientation of the nucleus with respect to the laboratory frame. This leads to the practical relationship, for $B(E2; I \rightarrow I - 2) := B_{I,I-2}$,

$$\frac{B_{I,I-2}}{B_{20}} = \frac{15I(I - 1)}{2(2I - 1)(2I + 1)} := f(I) \tag{1.6}$$

and the leading value, $B_{42}/B_{20} = 10/7 = 1.429$.

For electric quadrupole moments in ground-state bands of even–even nuclei there is a dependence on the orientation of the nucleus with respect to the laboratory frame. By convention,

$$\langle IMK|T(E2)|IMK\rangle = \langle I, M = I|T(E2)|I, M = I\rangle\langle I, K = 0|T(E2)|I, K = 0\rangle$$
$$= \langle II20|II\rangle\langle I020|I0\rangle(2I + 1)^{1/2}(5/16\pi)^{1/2}eQ_0, \tag{1.7}$$

whence

$$Q(I) = -\frac{I}{2I + 3}eQ_0.$$ (1.8)

A relationship between B_{20} and $Q(2)$ follows:

$$Q(2) = -\frac{2}{7}\sqrt{16\pi B_{20}}.$$ (1.9)

Equation (1.6) is applied to data in table 1.1 and figure 1.2. Equation (1.9) is applied to data in figure 1.3. The data shown are consistent with these simple relationships. Experimental uncertainties result in an unclear view of the limitations of these equations; but the averaged behaviour strongly supports nuclear rotation with a constant quadrupole moment as a function of increasing spin. While the relationships presented above are just stated here, the reader who wishes to explore how they are derived is directed to exercise 1-15.

An important implication of table 1.1 and figure 1.2 is that the model parameter Q_0 is consistent with being independent of I. In contrast, changes in the rotational energy parameter, A with respect to I, viz.

$$A = \frac{\Delta E_{I,I-2}}{4I - 2} = \frac{E_\gamma(I \rightarrow I - 2)}{4I - 2},$$ (1.10)

variations of which are shown in figure 1.4, indicate that something must be changing as the nucleus 'rotates'. But the constancy of Q_0 with increasing spin implies that it is not the deformation. This implication is not widely appreciated: many authors refer to centrifugal stretching of the moment of inertia, which is naturally based on a semi-classical view of the nucleus as a liquid drop. The view of the nucleus as a rotating liquid drop is evidently wrong. Phenomenological energy formulae do not reveal the origin of departures from equation (1.1). We address the interpretation of A in terms of a moment of inertia in section 1.4. We place the word 'rotates' in quotation marks because we will see that even the basic concept of rotation may not be correct.

Equation (1.1) emerges from the elementary model called the symmetric top. Details are presented in section 5.3 of [9]. It is the simplest possible view of quantum mechanical rotations and only assumes an axially symmetric shape with a plane of reflection symmetry at right angles to the symmetry axis for distribution of mass within the body. No internal degrees of freedom are assigned to the body; it is a rigid body. This simplicity appears valid for the distribution of electrical charge within the body, i.e., the distribution of the protons. We will progressively address this contradiction between the model parameters A and Q_0 as we proceed to look in depth at data.

Equations (1.5)–(1.9) depend on the concept of the K quantum number in association with the symmetric top. It is important to emphasize that this quantum number is a consequence of self-organization of a nuclear many-body system. These equations are the result of applying the Wigner–Eckart theorem to operators expressed in terms of the su(2) angular momentum algebra associated with nuclear

Table 1.1. Table of $E2$ transition strengths in W.u. for ground-state bands of deformed even–even rare earth nuclei, showing currently known values up to $I = 10$. The four right-hand columns give 'reduced' values, where for each nucleus the $4 \to 2$, $6 \to 4$, $8 \to 6$, and $10 \to 8$ values are scaled to the $B(E2; 2 \to 0) = B_{20}$ value and to the rotor spin factor $f(I)$, given by equation (1.6). The spin factor for the $4 \to 2$ transition is 1.429 and is noted in the column heading, with the other spin factors also given. Thus, for ^{154}Sm—245/176 × 1.429 = 0.97; with the generic form of the ratios expressed as $B_{I,I-2}/B_{20}f(I)$. The average value appearing in each column is given at the bottom, and the global average is 1.035, i.e., the rotor is realized at the 3.5% level. The data are taken from ENSDF (http://www.nndc.bnl.gov/ensdf/).

	$B(E2; I \to I - 2)$ (W.u.)					$B_{I,I-2}/B_{20}f(I)$			
						4	6	8	10
	$2 \to 0$	$4 \to 2$	$6 \to 4$	$8 \to 6$	$10 \to 8$	1.429	1.573	1.647	1.692
^{154}Sm	176^1	245^6	289^8	319^{17}	314^{16}	0.97^2	1.04^3	1.10^6	1.05^5
^{156}Gd	189^3	264^4	295^8	320^{17}	314^{14}	0.98^2	0.99^4	1.03^5	1.01^4
^{158}Gd	198^5	290^4	–	330^{30}	340^{30}	1.02^3	–	1.01^9	1.01^9
^{158}Dy	186^4	266^{15}	340^{40}	340^{70}	320^{50}	1.00^6	1.16^{24}	1.11^{23}	1.02^{16}
^{160}Dy	196^3	285^{11}	238^{13}	328^{30}	329^{15}	1.02^5	0.78^5	1.03^9	1.00^5
^{162}Dy	204^3	289^{12}	301^{17}	346^{17}	350^{23}	0.99^4	0.94^5	1.03^5	1.01^7
^{164}Dy	211^4	271^{11}	303^9	300^{13}	358^{18}	0.90^5	0.91^3	0.86^4	1.00^5
^{158}Er	129^9	186^6	246^8	298^{10}	250^{40}	1.01^7	1.21^9	1.40^{10}	1.15^{18}
^{160}Er	169^6	241^8	263^{15}	290^{90}_{60}	290^{70}	1.00^4	0.99^4	0.99^6	1.0^3
^{164}Er	205^5	260^{30}	–	343^{19}	353^{18}	0.88^{10}	–	1.01^6	1.01^5
^{166}Er	217^{15}	312^{11}	370^{20}	373^{14}	390^{17}	1.01^7	1.08^{10}	1.04^8	1.06^8
^{168}Er	213^4	319^9	424^{18}	354^{13}	308^{13}	1.05^4	1.27^5	1.01^4	0.85^4
^{162}Yb	135^4	210^9	191^{12}	250^{70}	180^{60}	1.09^4	0.90^7	1.1^3	0.8^3
^{164}Yb	162^5	259^9	276^{10}	320^{110}	300^{120}	1.12^4	1.08^5	1.2^4	1.1^4
^{166}Yb	191^{10}	272^9	291^{12}	320^{40}	310^{160}	1.00^6	0.97^6	1.02^{14}	1.0^5
^{172}Yb	212^2	301^{20}	320^{30}	400^{40}	375^{23}	0.99^7	0.96^{10}	1.15^{12}	1.05^6
^{174}Yb	201^7	280^9	370^{50}	388^{21}	335^{22}	0.97^4	1.17^{16}	1.17^7	0.99^5
^{176}Yb	183^7	270^{25}	298^{22}	300^{50}	320^{30}	1.03^9	1.04^8	1.00^{17}	1.03^{10}
^{166}Hf	128^7	202^7	221^{13}	280^{30}	–	1.10^5	1.10^9	1.33^{14}	–
^{168}Hf	154^7	244^{12}	285^{18}	350^{50}	370^{60}	1.17^7	1.18^{10}	1.4^2	1.4^2
^{170}Hf	182^7	263^4	306^{10}	344^{12}	375^{25}	1.01^5	1.07^5	1.15^6	1.22^8
^{178}Hf	160^3	–	219^{12}	237^6	257^8	–	0.87^6	0.90^3	0.95^4
^{180}Hf	155^2	230^{40}	219^{16}	245^{13}	238^{12}	1.0^2	0.90^7	0.96^5	0.91^5
^{170}W	124^3	179^{18}	189^{14}	190^{50}	170^{40}	1.01^{10}	0.97^7	0.9^2	0.8^2
^{172}W	171^{15}	245^{18}	260^{30}	290^{30}	270^{40}	1.00^9	0.97^{11}	1.03^{14}	0.93^{15}
^{174}W	135^9	235^{12}	410^{80}	240^{30}	160^{30}	1.22^9	1.9^4	1.08^{13}	0.70^{13}
^{182}W	136^2	196^{10}	201^{22}	209^{18}	203^{19}	1.01^5	0.94^9	0.93^8	0.88^8
^{184}W	120^2	166^5_9	181^6	185^5	220^{220}_{130}	0.97^3_5	0.96^4	0.94^4	1.10^6_{20}
^{186}W	111^2	144^{10}	187^{13}	178^{13}	151^{15}_{45}	0.91^6	1.07^7	0.97^7	0.8^1_2
Avg.						1.014	1.051	1.063	1.010

global average 1.035

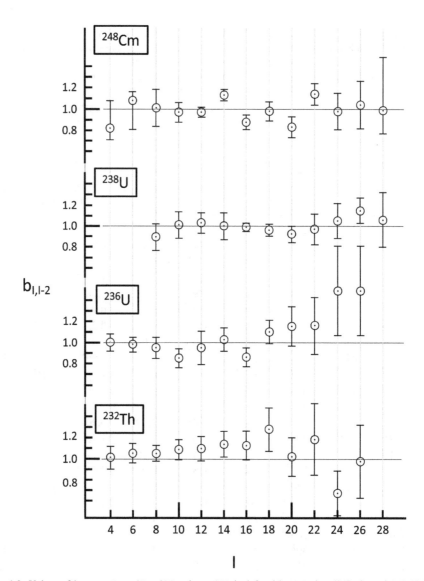

Figure 1.2. Values of $b_{I,I-2} := B_{I,I-2}/B_{20} f(I)$, where $f(I)$ is defined in equation (1.6), for spins 4–28 for the ground-state bands of all the actinide nuclei for which there are data. If the symmetric top is valid for nuclei, all the $b_{I,I-2}$ values should be unity.

rotation. Details of the Wigner–Eckart theorem and the su(2) algebraic structure of spin and angular momentum are presented in [10]. The band members are connected by the $E2$ operator and the common value of $K = 0$ for the band permits reduction of matrix elements, both transition and diagonal, to ratios of Clebsch–Gordan coefficients. The parameter, Q_0 is the so-called reduced matrix element of the Wigner–Eckart theorem. There is no *a priori* reason why this should emerge from a nuclear many-body system, but it appears to work well for the most strongly

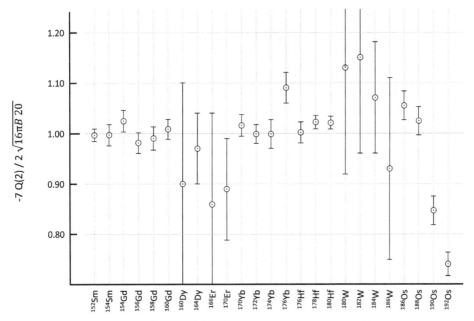

Figure 1.3. The ratio of $Q(2_1^+)$ to $\sqrt{B(E2;\, 2_1^+ \to 0_1^+)}$, reduced with respect to the rotor model scale factor (see equation (1.9) and text). Rigid symmetric top behaviour would correspond to this ratio being equal to unity. See later for a discussion of [186,188,190,192]Os.

deformed nuclei. An introduction to the role of quantum numbers and the Wigner–Eckart theorem, in arriving at the simple relationships embodied in equations (1.5)–(1.9), is given in the next section.

1.2 Quantum numbers and the Wigner–Eckart theorem for nuclear rotation

The quantum numbers that label nuclear rotational states emerge from the quantum mechanics of the symmetric top model. These quantum numbers are depicted in figure 1.1. The governing quantum number of the state of a nucleus, in this context, is the total spin, I. There is a second quantum number contingent upon the value of I, its directional component. Recall that only a single directional component of I is allowed due to quantum uncertainty. Thus, one speaks of the 'cone of indeterminacy' in the quantum theory of angular momentum. This is depicted in figure 1.5. With respect to deformed nuclei, and specifically the symmetric top model, there are two independent frames of reference involved: the body-fixed frame and the laboratory-fixed frame. With respect to these two frames of reference, directional quantum numbers in the body-fixed frame and in the laboratory-fixed frame are defined, K and M, respectively.

The defining of two frames of reference with respect to nuclear rotation is mandatory. It is impossible to define a precise orientation of a deformed nucleus. Indeed, this is impossible for any finite many-body quantum system when any type

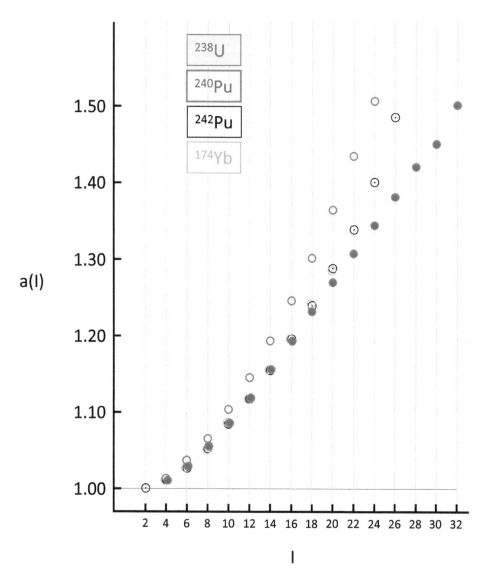

Figure 1.4. Values of the scaled rotational energy parameter for selected nuclei. These are defined using equation (1.10), expressed as $a(I):=[E_\gamma(I \to I - 2)/(4I - 2)]/[E_\gamma(2 \to 0)/6]$. A rigid rotor would result in $a(I) = 1.000$ for all values of I. Experimental uncertainties for the input energies are too small to be shown. Note the values of $a(I)$ for ^{242}Pu and ^{174}Yb are almost indistinguishable for all spin values: this is discussed further in section 1.3, including details of the uncertainty in their energies.

of 'deformation' of the system arises. For deformed nuclei, the two frames of reference can be viewed, in pictorial terms, as shown in figure 1.6: we refer to this view as the 'hyper-cone of indeterminacy'. This view of the atomic nucleus underlines a fundamental limitation to our language for discussing such deformed systems: we make observations in the laboratory frame but we are describing the quantum mechanics in the body frame.

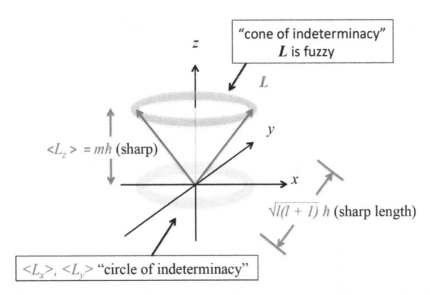

Figure 1.5. The 'cone of indeterminacy' for angular momentum in quantum mechanics. It is impossible to define more than one directional component of angular momentum, i.e. only one directional component is 'sharp'. Thus, in physical space, the other two (Cartesian) components can only be defined to within a circle. This results in the useful depiction of the quantum mechanical uncertainty using a cone of height m and side $\sqrt{l(l+1)}$, in units of $\hbar$.

We could formulate nuclear properties entirely in laboratory-frame coordinates. The equations would be intractable. (Consider, if we attempted this when describing physical processes on the surface of the Earth as viewed from Space.) Describing physical processes occurring in systems that are rotating is enormously simplified by formulating the description in the body-fixed (rotating) frame. Note that, when this is done, so-called 'Coriolis' and 'centrifugal' effects are encountered: these are most easily viewed from the laboratory frame. (Consider, understanding a rotating air mass in a storm system and the flow of air towards the North Pole and the consequent acquisition of easterly motion, as viewed from Space looking down from above the North Pole.) We will address Coriolis and centrifugal effects in nuclei in due course (with some major surprises). (If we expressed properties of nuclei in a laboratory-frame of reference, from all the intractable equations we would find the emergence of some very simple relationships: these relationships would be difficult, even impossible to understand; but would be transparent, even trivial when expressed in a body frame of reference.)

The observational basis of nuclear spectroscopy is the determination of expectation values of quantities such as energies, spins, quadrupole moments, and transition probabilities (radiation intensities). These are formally expressed as 'diagonal' matrix elements of operators for expectation values and 'transition' matrix elements of operators for transition probabilities. In a quantum-mechanical model description, one works in a basis of energy eigenstates which possess simultaneous quantum numbers, in the present context angular momentum quantum numbers. Thus, the quantum states of the symmetric top model, in the

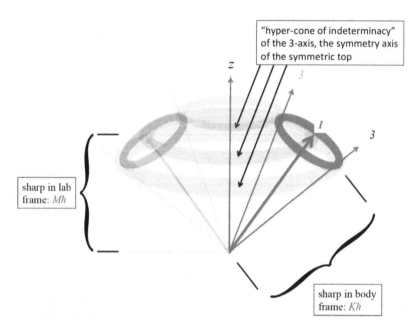

Figure 1.6. The 'hyper-cone of indeterminacy' for the quantum mechanical axially symmetric rotor. The body-fixed frame and the laboratory frame are not in a fixed relationship to each other: they are connected by the total spin I possessing a fixed projection $M\hbar$ on the laboratory frame z-axis and a fixed projection $K\hbar$ on the body frame 3-axis. In consequence, the 3-axis has an uncertainty that involves two cones of uncertainty from which the figure endeavours to depict the full degree of uncertainty possessed by the quantum axially symmetric rotor.

body-fixed frame, are defined by energy, I, and K, viz. $|E, I, K\rangle$. For present purposes, just three quantum numbers[4] define the entirety of the quantum mechanics of the model. Most importantly, they define the quantum mechanical basis within which all properties of the system are formulated.

Physical processes in quantum mechanical systems are described by operators. To understand how operators work (operate) within the system, one must know how the operators act on the basis states. This is variously formulated in quantum theory using differential operators acting on functions, matrices acting on vectors, or algebraic formulations. For present purposes, an algebraic formulation is by far superior in conciseness and ease of use; but it is the most abstract. We remind the reader that any formulation of quantum mechanics involves the preciseness of mathematics, with the mandate that the description cannot imply more information than is physically achievable by measurement. Herein lie the features of quantum theory that defy everyday logic.

Operators act on basis states to produce other basis states, or the same basis state: if the state is another basis state, a transition has occurred; if the state is the same basis

---

[4] There is a parity quantum number that is contingent on I, also on internal symmetries of the body: we do not discuss this here as it does not play any role; but some details are presented in [9], and later in this volume.

state, that defines an observable of the system. Transitions are also observable, but there is a profoundly subtle aspect to transitions in quantum systems: namely, we learn about quantum systems by a focus on energy eigenstates, so-called stationary states. Energy eigenstates 'do not do anything' within the system. For a transition to occur, the system must be coupled (interact) with another system. For nuclei this could be by collision with another nucleus. It can also be by coupling to what we call the electromagnetic vacuum. This is always present; but when a nucleus is in an excited state (an energy eigenstate), e.g. after having been produced by a nuclear collision, the excited nuclear state is not just under the control of the Hamiltonian for the nucleus; it is also coupled to the Hamiltonian for the electromagnetic vacuum. Thus, the nucleus in its excited state, while in an energy eigenstate of the nuclear Hamiltonian, is not in an energy eigenstate of the total Hamiltonian: it emits electromagnetic radiation and undergoes a transition to a new energy eigenstate of the nuclear Hamiltonian.

One might ask: 'how come the nucleus ends up in an eigenstate of the nuclear Hamiltonian and not that of the total (nuclear + electromagnetic, field) Hamiltonian?' The answer is that the coupling between the two is weak. If the coupling was strong, the states of the nucleus would not be observed as sharply defined energies: such states are observed in nuclei as resonances. These resonances have broad energy distributions because they are unbound with respect to neutron or proton emission; thus, one must consider such unbound states in the full basis of description and these unbound states have continuous energy distributions. In fact, due to coupling to the em vacuum, bound excited states have energy 'widths': these are expressed as $\Delta E = \hbar/\tau$, where τ is the mean lifetime of the quantum state. For electromagnetic decay, these widths are generally far smaller than the precision with which we can measure the energies of excited states in nuclei (detector energy resolution). But such widths are seen in laser spectroscopy of excited states in atoms. Energy widths observed in meson and baryon spectroscopy are nearly all enormous, for quite different reasons than in nuclei[5].

The handling of operators and how they act on basis states is enormously simplified when symmetry is possessed by the physical system. The initial and final states in any process must possess this symmetry for fast decay modes (the dominant modes). As such, we classify operators by such symmetry. An operator may or may

---

[5] Baryons and mesons (hadrons) exhibit excited states due to their internal quark and anti-quark degrees of freedom (and probably their gluon degrees of freedom). But quarks (and gluons) are absolutely bound. The widths come from the strong coupling of these excitation degrees of freedom to decay processes such as the emission (creation) of, e.g. π^0 or $\pi^+\pi^-$ pairs (a certain analogy to the internal-conversion and internal-pair decay modes in nuclei exists here). But there are a few excited states in baryon and meson systems that only have small energy widths: these are states that can only decay by the weak interaction. The historical sensation was the so-called J/psi particle, a meson formed of a charm-anti-charm quark pair. This 'charmonium' system has excited states that predominantly decay by gamma-ray emission, i.e., through their coupling to the em vacuum: they also have relatively narrow widths. One can take this line of thinking back into atomic systems with the example of the 'atom' formed by an electron–positron pair, positronium. Positronium has singlet and triplet lowest-energy states. Unlike hydrogen, positronium has a decay channel: annihilation. The singlet and triplet lowest-energy states have different widths due to symmetry with respect to decay into two or three photons, following the annihilation process. This is a rich arena for understanding time-dependent processes related to finite bound quantum systems.

not possess this symmetry. If it does possess the symmetry, our job is almost done (see below); if it does not, we expand the operator in basis components defined by the symmetry. To put this into practical terms, electromagnetic decays of the nucleus can be expressed in terms of the spherical harmonics. We refer to this expansion as multipoles of the radiative process (coupling to the em field). While such an expansion may contain many terms (multipoles), only one or a few components in the expansion dominate. The reason is that spherical harmonics are defined also by an angular momentum-type label, and photon emission from nuclei is highly restricted by angular momentum such that a spin change of more than one or two units is very improbable (recall that the spin of a photon is $1\hbar$). This is the basis of the so-called 'Weisskopf' estimate for electromagnetic transition strengths in nuclei.

Spherical harmonics are labelled by indices that match the mathematical structure of the angular momentum theory of quantum systems. We could say that they are 'symmetry-adapted'. Thus, spherical harmonics appear as representations in a wave function description of electronic states in atoms. They are not simply adapted to a wave function description of nuclear rotations (because of the hyper-cone of indeterminacy, they are too sharp, i.e., they would imply more information than we possess[6]). But all we really need from the concept of the spherical harmonics is that of multipolarity, their indices. This takes us into the concept of the spherical tensor structure of operators, and hence to the Wigner–Eckart theorem.

Spherical tensors are mathematical entities just like scalars and vectors. One can manipulate them: add them together; operate on them, e.g. rotate a vector. But there is an over-riding constraint: a vector cannot be added to a scalar. The underlying mathematical generalization is the concept of tensorial character. A scalar has just one 'component', which formalizes the concept of 'number'. A vector in the space in which we live has three components, say (x, y, z). A four-dimensional space has vectors with four components and so on. Spherical tensors are indexed by two numbers, conventionally by $\{l, m\}$ or $\{\lambda, \mu\}$: for a given λ, $\mu = +\lambda, +\lambda - 1, \ldots, -\lambda$, i.e. $2\lambda + 1$ components. Thus, for $\lambda = 1$, there are three components, and such a spherical tensor is said to be isomorphic (identical in form) to a cartesian vector in three-dimensional space: this can be expressed, for $T_\mu^{(\lambda)}$, with $\lambda = 1$, as $T_{+1}^{(1)} = x + iy$, $T_{-1}^{(1)} = x - iy$, $T_0^{(1)} = z$. Operators can be expressed as spherical tensors, angular momentum states have identical spherical tensor structure, so there are rules for combining them.

---

[6] Rotational wave functions or Wigner-D functions are quite complicated. They can be expressed in their full detail as 'arrays' of products of pairs of spherical harmonics, sometimes appearing in a matrix form. The reason is that two directional indices must be used for quantum mechanical rotations, the earlier defined K and M quantum numbers. This double indexing, and doubly expressed uncertainty, is because the orientation of the laboratory-fixed frame of reference and the body-fixed frame of reference has an inherent quantum mechanical uncertainty, as shown in figure 1.6. Thus, we cannot use single spherical harmonics: the Wigner-D functions—arrays of products of pairs of spherical harmonics, one labelled by K and one labelled by M— express this uncertainty in precise mathematical terms. They are usually written $D_{MK}^I(\theta, \phi)$, where the argument of the function involves the spherical polar angles describing the orientation of the body-fixed frame with reference to the laboratory-fixed frame, indexed by the total spin I. We do not use the Wigner-D functions herein.

The quantum theory of angular momentum, whether applied to nuclei or atoms, or any other finite many-body quantum system provides rules for coupling states of angular momentum together, e.g. coupling a spin to an orbital angular momentum or coupling two particle spins together to obtain a resultant total spin. So, one can 'couple' a spherical tensor operator to an angular momentum state: this is how an operator can be viewed as acting on a state. One obtains a final state in a physical process, which can be viewed as coupling the 'spin' of the operator to the spin of the state. The rules for such coupling are governed by Clebsch–Gordan, CG, coefficients.

For the action of operators on states to produce new states (transitions within the system) this enables such processes to be formulated in terms of CG coefficients. This is expressed as

$$\langle I_f K_f | T_\mu^{(\lambda)} | I_i K_i \rangle = \langle I_f K_f \lambda \mu | I_i K_i \rangle \langle I_f || T^{(\lambda)} || I_i \rangle / (2I_i + 1), \tag{1.11}$$

where the term on the left is the matrix element describing the physical process, the first term on the right is a CG coefficient, and the 'double-barred' expression on the right is called the 'reduced' matrix element (the denominator on the right is a conventional factor). This is the statement of the Wigner–Eckart theorem. It provides an enormous reduction in computational labour; often just involving a look-up of the CG coefficient. Even, if the CG coefficient is zero, this leads to the concept of a 'forbidden' transition.

What is remarkable about the rotational states of nuclei is that in equation (1.11), the reduced matrix element is a common numerical factor for many matrix elements. Specifically, for the $K = 0$ rotational band built on the ground state of a doubly even nucleus, all transition and all diagonal matrix elements reduce within experimental error to a single number (multiplied by a CG coefficient), which we define to be Q_0: the so-called intrinsic quadrupole moment of the nucleus. This is arguably the best manifestation of the Wigner–Eckart theorem in the entire domain of quantum theory. This 'reduction' depends on the validity of the quantum number K, which is a model quantum number. Note that in equation (1.11), on the right-hand side, the reduced matrix element is independent of K: one could say that the Wigner–Eckart theorem has 'factored-out' K and isolated it in the CG coefficient.

Note that nowhere does one need to formulate the quantum mechanics of the electromagnetic field in the above details. One only needs to know that precise amounts of energy can be exchanged between the nucleus and the em field (with allowance for recoil energy of the nucleus when high-precision measurements are used); and that rates of decay, probabilities of exchange of energy, have a dependence on the angular momentum change and the parity change. For the quadrupole degree of freedom, the spin change is two, with no parity change. Again, we note the role of the Weisskopf estimates; herein combined with the observation that collective degrees of freedom exhibit large enhancements over these estimates (large $B(E2)$ values).

The full details of the quantum mechanics underlying this section are given in the earlier volumes in this series: (chapter 9 of [11]) an introduction to time-dependent quantum mechanics, (chapter 11 of [11]) an introduction to the algebraic structure of angular momentum and spin, (chapter 1 of [10]) spherical harmonics, (chapter 2 of [10]) coupling of angular momenta and spins, (chapter 3 of [10]) tensor structure of operators and the Wigner–Eckart theorem, (chapter 8 of [10]) time-dependent perturbation theory, (chapter 9 of [10]) the electromagnetic field in quantum mechanics. Note that handling the multipole expansion of the electromagnetic field is not yet developed in the Series (only the electric dipole approximation was made in chapter 9 of [10]); higher multipole terms will be needed, e.g. for the theory of angular correlations between sequential radiative decay steps.

1.3 Odd nuclei: energies and *E*2 properties

We can immediately look more deeply at the issue of rotations in nuclei by inspecting data for odd-mass nuclei. The above equations are for ground-state bands of even–even nuclei, i.e., for $K = 0$. For general K values these equations are modified by replacement of the CG coefficients, viz. $\langle I020|I - 2, 0\rangle \rightarrow \langle IK20|I - 2, K\rangle$ in equation (1.5) and $\langle I020|I0\rangle \rightarrow \langle IK20|IK\rangle$ in equation (1.7). For example, this leads to the relationship (cf equation (1.7))

$$Q(I, K) = \frac{\left[3K^2 - I(I + 1)\right]eQ_0}{(I + 1)(2I + 3)}.$$

(1.12)

Figures 1.7(a)–(f) show values of the parameter Q_0 extracted from data for selected odd-mass nuclei using equation (1.4) and (cf equation (1.5))

$$\langle I, K|T(E2)|I - 2, K\rangle = (2I + 1)^{1/2}(5/16\pi)^{1/2}\langle IK20|I - 2, K\rangle eQ_0.$$

(1.13)

The model interpretation is again consistent with Q_0 being independent of I. Comparison of Q_0 values for odd-N nuclei with their even–even neighbours is shown in figure 1.8; a similar comparison for odd-Lu isotopes with even-Yb and even-Hf neighbouring isotopes is shown in figure 1.9. Figure 1.10 shows Q_0 values in the actinide region. The pattern is consistent with a smooth variation in Q_0 as a function of mass number. While all the even–even nuclei involve $K = 0$ states, the odd nuclei have a range of K values; but Q_0 systematics are smooth and illustrate the independence of this quantity with respect to K. This is a practical illustration of the way that the Wigner–Eckart theorem works, beyond the data showing that Q_0 values are consistent with a single value in a rotational band.

Although energies of states in rotational nuclei do not conform exactly to equation (1.1), one can inspect energy patterns in neighboring nuclei to seek similarities and differences. The comparison of odd and even–even neighbours is made for γ-ray transition energies for selected nuclei in figures 1.11–1.13. In some cases, the differences implied for the rotational energy parameter(s) are ~0.1%. In general, differences in energies between odd-mass nuclei and even–even nuclei depend on a so-called 'rotation-particle coupling' term. From the rotational energy Hamiltonian,

$$H = A\mathbf{R}^2,$$

(1.14)

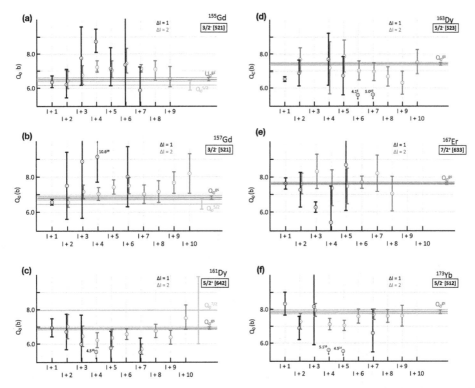

Figure 1.7. (a–f) Intrinsic quadrupole moments Q_0 in barns versus spin for ground-state bands of selected odd-N nuclei in the rare earth region, extracted using the axially symmetric rotor model, cf equations (1.4), (1.12), (1.13). These values are based on $B(E2)$ values given in ENSDF (except for ^{157}Gd where failure to use an erratum [12] results in all the ENSDF values for states with spin above 7/2 being wrong). The values of Q_0 and their uncertainties deduced from the ground-state spectroscopic quadrupole moments are shown in red, with extension across the entire range of spins of excited states so that they form a visual base reference for each nucleus. Where excited state quadrupole moments have been measured, in $^{155,\ 157}$ Gd, ^{161}Dy, these are shown in green and the centroid of the value again extends across the entire range of spins of excited states, but the uncertainties are localized using standard error bar notation. The spectroscopic quadrupole moment values are taken from ENSDF and [13]. The most precise values of Q_0 are for the ground states, which result from the spectroscopic quadrupole moments. Excited band members appear to be consistent with no changes in Q_0 as spin increases, but the precision is insufficient to make strong statements regarding constant Q_0 values for any of these bands.

where **R** is the collective angular momentum of the core and A is the rotational energy parameter, cf equation (1.1) and figure 1.1. Then, defining

$$\mathbf{I} := \mathbf{R} + \mathbf{j}, \tag{1.15}$$

where **I** is the total spin of the nucleus and **j** is the spin of the odd nucleon,

$$H = A(\mathbf{I} - \mathbf{j}) \cdot (\mathbf{I} - \mathbf{j}) = A\mathbf{I}^2 - 2A\mathbf{I} \cdot \mathbf{j} + A\mathbf{j} \cdot \mathbf{j}. \tag{1.16}$$

On quantization, this leads to

$$E = AI(I + 1) - 2A\langle \mathbf{I} \cdot \mathbf{j} \rangle + A\langle \mathbf{j} \cdot \mathbf{j} \rangle, \tag{1.17}$$

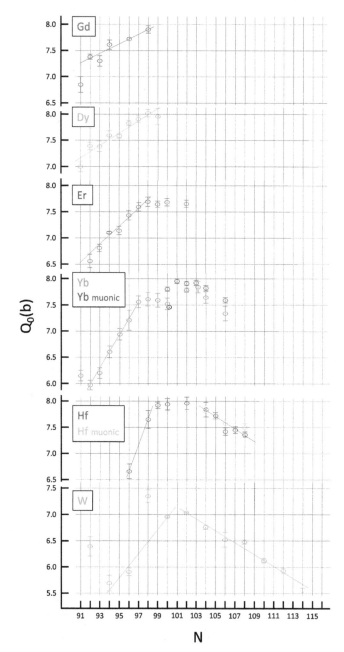

Figure 1.8. Comparison of Q_0 values in barns for ground states in odd-N nuclei and 2_1^+ states in even–even neighboring nuclei, Gd–Hf. Sloping lines are to suggest a more rapid onset of deformation in going from Gd to Yb. The values are computed using data taken from ENSDF and [13]. The value of Q_0 for ^{161}Dy is taken from the fit to the muonic hyperfine structure allowing for K mixing [14]. The two sets of values shown for some of the Yb and Hf isotopes correspond to evaluated data in ENSDF and data from muonic x-ray hyperfine structure: specifically, data from [15] and data from [16]. For ^{170}Yb, three values are shown: $Q_0^{\text{Coul.}} = 7.4617$ b [black], $Q_0^{\text{muonic}} = 7.804$ b [blue], and $Q_0^{\text{ENSDF}} = 7.5211$ b [orange].

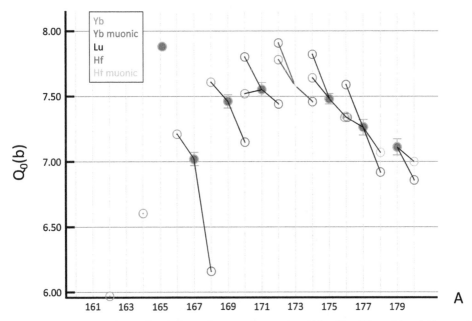

Figure 1.9. Comparison of Q_0 values in barns for ground states in odd-Lu ($Z = 71$) nuclei and 2_1^+ states in even–even neighboring Yb ($Z = 70$) and Hf ($Z = 72$) nuclei. Uncertainties are shown only for the Lu isotopes. Lines connect the Yb–Lu–Hf isotones. The values are computed using data taken from ENSDF and [13].

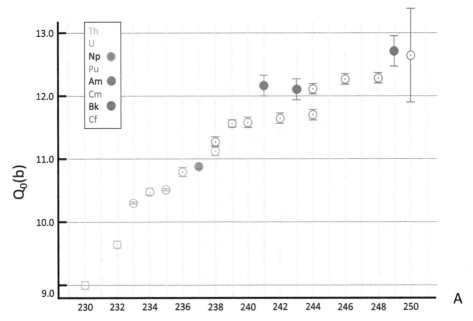

Figure 1.10. Comparison of Q_0 values in barns for ground states in actinide nuclei. The values are computed using data taken from ENSDF and [13].

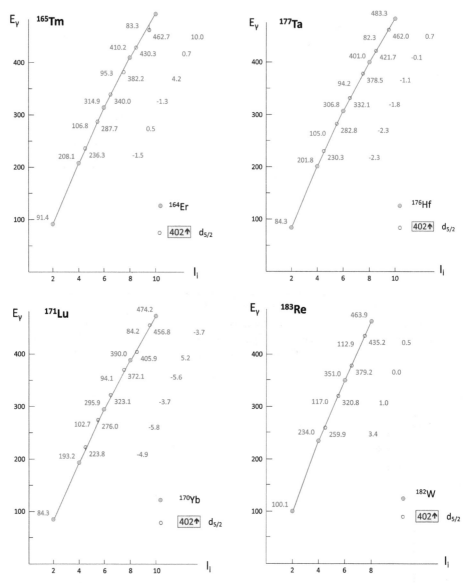

Figure 1.11. (a–d) Plot of γ-ray transition energy, E_γ in keV versus spin of the initial state, I_i for rotational bands built on the Nilsson state $5/2^+$ [402] for selected odd-Z nuclei spanning $N = 96$ to 108, compared to neighbouring $A - 1$ even–even 'core' nuclei. All the transitions have $\Delta I = 2$ and the plots are limited to the lowest spin band members. Experimental values are given in red; energy differences (even mass, self-evident) or interpolated energy differences (e.g. $106.8 \times 0.25 + 208.1 - 236.3 = -1.5$) are given in blue. Thus, for the odd-mass nuclei, if the difference is negative the data point lies above the even–even 'trajectory': this convention is in recognition of the negative sign in the RPC term in equation (1.17). Details are discussed in the text. The data are taken from ENSDF.

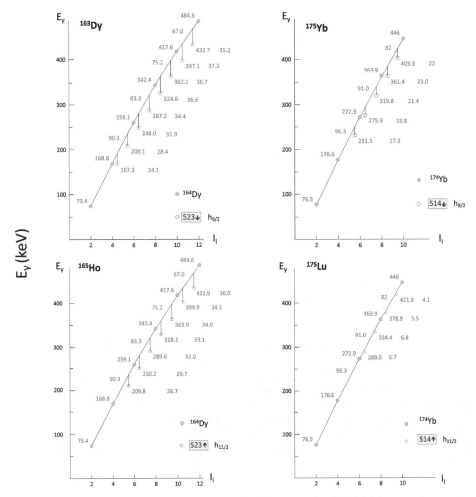

Figure 1.12. (a–d) Plot of γ-ray transition energy, E_γ in keV versus spin of the initial state, I_i for rotational bands built on the Nilsson states 5/2⁻ [523] (¹⁶³Dy), 7/2⁻ [523] (¹⁶⁵Ho) and 7/2⁻ [514] (¹⁷⁵Yb), 9/2⁻ [514] (¹⁷⁵Lu). Note the common reference cores; also note that these are unique-parity configurations involving $l = 5$, $j = 9/2 = 5 - 1/2$ (odd-neutron nuclei with 523↓ and 514↓) and $l = 5$, $j = 11/2 = 5 + 1/2$ (odd-proton nuclei with 523↑ and 514↑). For other details, see the caption to figure 1.11 and discussion in the text. The data are taken from ENSDF.

where the first term is familiar, cf equation (1.1); the second term is called the 'rotational-particle coupling', RPC, term (also, sometimes, it is called the 'Coriolis' term) and the third term is called the 'recoil' term. These are expressed as expectation values with respect to specific states in a rotational band and are handled shortly. (The appearance of expectation values is because j is not a good quantum number in a deformed mean field, i.e. j values are mixed; this is handled later.) It is important to note that the second term is linear in I and the third term is independent of I. Thus, energy *differences*, notably $E_\gamma(I \rightarrow I - 2)$ can be approximated by

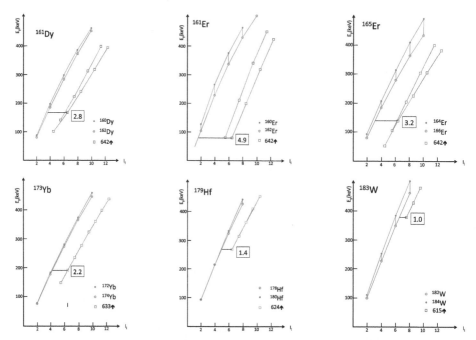

Figure 1.13. (a–f) Plot of γ-ray transition energy, E_γ in keV versus spin of the initial state, I_i for rotational bands built on Nilsson states originating from the $i_{13/2}$ configuration. The 'alignment spins' for the $13/2 \to 9/2$ transitions (and $15/2 \to 11/2$ in ^{183}W) are given in the boxes. Both the $A-1$ and the $A+1$ core nuclei are shown; the alignment spins are relative to the trajectory of the core with the lower set of energies. The pattern suggests that alignment decreases with increasing Ω, i.e. 633↑ ($\Omega = 7/2$), 624↑ ($\Omega = 9/2$), 615↑ ($\Omega = 11/2$), cf ^{173}Yb, ^{179}Hf, ^{183}W, and with increasing deformation, cf ^{161}Er/^{161}Dy and ^{161}Er/^{165}Er (note configurations are all 642↑ ($\Omega = 5/2$) and line slopes are ^{161}Er > ^{161}Dy, ^{161}Er > ^{165}Er, where line slope is fixed by rotational energy constant which is inversely proportional to the moment of inertia). Other details are explained in the caption to figure 1.11 and in the text. The data are taken from ENSDF.

$$E_\gamma(I \to I - 2) = A(4I - 2) - 2A(2j \cos \theta_{Ij}), \qquad (1.18)$$

where θ_{Ij} can be viewed as the semi-classical angle between the average j of the unpaired nucleon and the total nuclear spin, I. It is evident from figures 1.11–1.13 that the second term in equation (1.18) is indeed generally independent of I. In many nuclei it is almost zero. The largest non-zero occurrences of the second term in equation (1.18), which are for configurations originating in high j-value spherical shell model structures, and are termed 'alignment' energies, are shown in figures 1.14(a)–(c).

The rotation-particle coupling term in the Hamiltonian generally has a small influence on electromagnetic properties in odd-mass nuclei. However, for $K = 1/2$ bands it can be significant. The contribution of this term to energies is discussed in chapter 3 of [9] (especially figures 3.11 and 3.12; and cf equation (3.16) therein). For $K = 1/2$ bands and $E2$ matrix elements

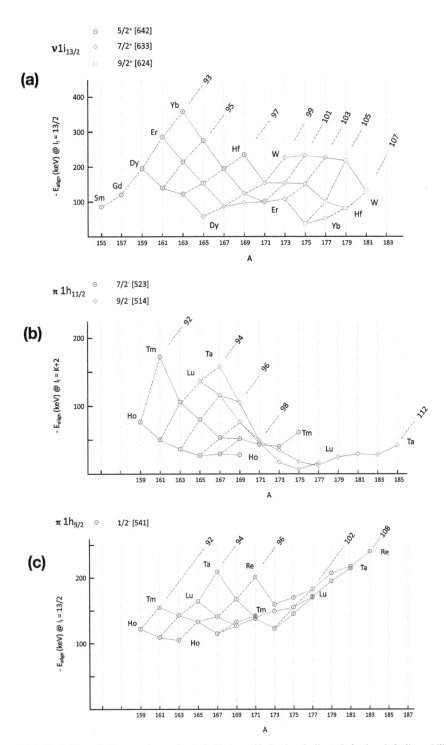

Figure 1.14. (a–c) Plot of 'alignment' energies in keV at specified spins (indicated) for bands built on Nilsson configurations from: (a) the $1i_{13/2}$ configuration; (b) the $1h_{11/2}$ configuration; (c) the $1h_{9/2}$ configuration. See the text for further details.

$$\langle I_i, 1/2|\boldsymbol{T}(\boldsymbol{E}2)|I_f, 1/2\rangle = (2I_i + 1)^{1/2}(5/16\pi)^{1/2}\langle I_i, 1/2, 2, 0|I_f, 1/2\rangle eQ_0X, \quad (1.19)$$

where

$$\begin{aligned}
X = 1 + \zeta[(I_i - 1/2)^{1/2}(I_i + 3/2)^{1/2}\langle I_i, 3/2, 2, -1|I_f, 1/2\rangle \\
+ (I_i + 1/2)\langle I_i, -1/2, 2, +1|I_f, 1/2\rangle],
\end{aligned} \quad (1.20)$$

ζ is a parameter, and the expressions '$\langle\ |\ \rangle$' are CG coefficients: the derivation of equation (1.20) is explained later in the series. An example of high-precision data for $E2$ properties is available for the nucleus ^{239}Pu and this is illustrated in figure 1.15. Note that the modification of the $E2$ matrix elements by the rotation-particle coupling term produces a convergence between the model and the experimental data. Thus, a simple correction to the zeroth-order manifestation of the model leads to an improved description of the data.

We make an important philosophical point in view of the agreement between data and the simple rotational model being used herein: namely, a symmetric top with an $\boldsymbol{I} \cdot \boldsymbol{j}$ particle-rotor coupling. When agreement between a model and data is 'good' in zeroth order and converges to 'very good' when the first-order model correction is

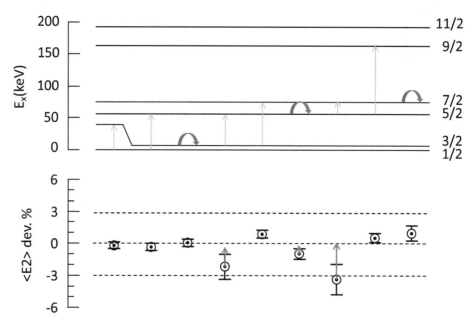

Figure 1.15. Deviation from the symmetric rigid rotor of experimental $E2$ matrix elements (in eb), determined using muonic x-ray spectroscopy, for the $K = 1/2$ ground-state band in ^{239}Pu. The red arrows show the changes produced by including RPC effects in the symmetric rotor description (other matrix elements change by less than 0.4%). The data are taken from [17]. The upper part of the figure shows diagonal $E2$ matrix elements as blue (curved) arrows and off-diagonal $E2$ matrix elements as green arrows. The zeroth-order fit uses a single parameter, $Q_0 = 11.592$ b; the fit with RPC effects uses two parameters, $Q_0 = 11.583$ b and $\zeta = -1.8 \times 10^{-3}$, see text, equations (1.13), (1.19) and (1.20). The figure design is a copy of one appearing in the above-cited paper.

made, one must consider this to be a valid fundamental feature of the many-body system. Even if the example is rare, it is unlikely to be an accidental occurrence.

We emphasize, again, the evident validity of the K quantum number, which in odd nuclei is not zero. Indeed, one can explore the implied values of K manifested in the $E2$ matrix elements by 'reverse engineering' the reduction process involved in the Wigner–Eckart theorem, i.e. imposing a fixed value of Q_0 for a rotational band and extracting the reduction factors—the CG coefficients. We extend this perspective to transitions between bands later.

Energies for $K = 1/2$ bands and bands in odd–odd nuclei are handled later. There are almost no $E2$ data for bands over extended ranges of spin in deformed odd–odd nuclei and excited bands in deformed odd nuclei. Thus, for multiple rotational bands, in a single nucleus, there is a lack of data for exploring whether Q_0 is a universal feature for multiple bands, or if each band is characterized by a different Q_0 value. This lack of data is due to the difficulty of obtaining a 'clean' view of the population and subsequent de-excitation of a given excited state in a nucleus. Unless the population of the given state is simple, it becomes (nearly) impossible to deconvolute the feeding 'history' of the state. The feeding history must be determined to allow for delays in feeding: this is handled using the so-called Bateman equations.

1.4 A wider look at rotation in nuclei: energies and moments of inertia

One can look at energy patterns beyond equation (1.1) via scaling of rotational band energies in even nuclei by $E(2_1^+)$. This received a limited inspection in figure 1.4. But in the finer details, a remarkable feature emerges, as presented in table 1.2 for the comparison of ^{174}Yb and ^{242}Pu: when scaled, the transition energies between states in the ground-state rotational bands are all identical to within a few parts in a thousand, independent of spin. Further, when considering experimental uncertainties, there is the possibility that these scaled energies are even more similar. We emphasize, these are supposedly complex many-body quantum systems with 174 and 242 bodies, respectively, with very different 'orbital occupancies' for the constituent nucleons. Indeed, similar patterns emerge when numerous ground-state bands are scaled in this manner. We note that these scaled energies are closer than any available phenomenological descriptions. There is no known explanation of this, i.e. at the level of the behaviour of nucleons in the nucleus.

The simplest interpretation of energy patterns for bands in nuclei is that the nuclei are deformed, and the energies of band members are characterized by a moment of inertia parameter. From knowledge of the mass, size and deformation of a given nucleus, a classical moment of inertia can be calculated. For a nucleus with

$$R(\theta, \phi) = R[1 + g + \beta Y_{20}(\theta, \phi)], \tag{1.21}$$

$g = -\beta^2/4\pi$ (volume conservation), we obtain

$$Q_0 = 3/\sqrt{(5\pi)}\, ZeR^2\beta(1 + \sqrt{(5/\pi)}\beta/8 + 5\beta^2/8\pi - (5/\pi)^{3/2}\beta^3/192 + \cdots), \tag{1.22}$$

Table 1.2. Comparison of ground-state band transition energies for ^{242}Pu and ^{174}Yb. The data are taken from ENSDF.

I_i	$E(^{242}$Pu$)$ (keV)	$E(^{174}$Yb$)$ (keV) $\times 0.5824$	$E(^{174}$Yb$)$ (keV)	% dev.
2	44.54^2	44.54 [norm.]	76.471^1	–
4	102.8^1	102.9	176.645^2	+0.098
6	159.0^1	158.9	272.918^6	−0.063
8	211.7^4	211.8	363.64^5	+0.047
10	260.5^6	260.4^6	447.2^{10}	−0.038
12	305.8^8	305.4^8	524.4^{13}	−0.131
14	347.3^{10}	347.1^{10}	595.9^{17}	−0.058
16	385.0^{11}	384.4^{11}	660^2	−0.156
18	419.3^{12}	418.7^{17}	719^3	−0.143
20	450.2^{13}	450.8^{29}	774^5	+0.133

−0.035 (avg.)

whence

$$91.7436 Q_0 / Z A^{2/3} = \beta + 0.157\,70\beta^2 + 0.198\,94\beta^3 - 0.010\,457\beta^4. \qquad (1.23)$$

Then, using

$$\mathscr{I}_{\text{rigid}} = 2/5 A M R^2 \left\{ 1 + \sqrt{(5/16\pi)}\beta + 0.44\beta^2 \right\}, \qquad (1.24)$$

where M is the mass of the nucleon and $R = 1.2 A^{1/3}$ fm,

$$\mathscr{I}_{\text{rigid}} = 9.6405 \times 10^{-58} A^{5/3} \left\{ 1 + 0.3154\beta + 0.44\beta^2 \right\} \qquad (1.25)$$

in kg m^2. From equation (1.1), with $A = \hbar^2/2\mathscr{I}_{\text{expt}}$, where recall that $1\ \text{J} = 6.241\,509 \times 10^{18}$ eV,

$$\mathscr{I}_{\text{expt}} = 2.0824 \times 10^{-52} / E(2_1^+\ \text{keV}) \qquad (1.26)$$

in kg m^2. Values for $\mathscr{I}_{\text{rigid}}$ and $\mathscr{I}_{\text{expt}}$ are compared below in table 1.3.

There are rotational bands with near constant energy differences for transition energies. Some superdeformed bands exhibit this: one of the best examples is shown in figure 1.16. This depicts a superdeformed band in ^{152}Dy via observed gamma-ray transition energies. The notable feature is the extraordinary constancy of the differences between these gamma-ray energies: an enhanced view is depicted at the bottom of the figure.

The consequences of equations (1.21)–(1.26) for ground-state bands in ^{174}Yb and ^{242}Pu, and for the superdeformed band in ^{152}Dy, are given in table 1.3.

The results manifested in table 1.3 are profound with respect to the physics of nuclear rotation. It means that nuclei probably approach rigid rotation asymptotically as deformation increases; and the rigid rotation limit is manifestly reached in

Table 1.3. Moments of inertia for ground-state rotational bands of a rare earth and an actinide nucleus and for a superdeformed rotational band. The moments of inertia are given in units of kg m^2 ×10^{-54} and the 2$^+$ energies are given in keV. Note that the 2$^+$ energy given for ^{152}Dy is an estimate based on extrapolation of values observed in association with spins >24, cf figure 1.16 and equation (1.31). See the text for other details and remarks.

	Z	Q_0	$A^{2/3}$	β	$\mathscr{I}_{rigid}$	$E(2_1^+)$	$\mathscr{I}_{expt}$	$\dfrac{\mathscr{I}_{expt}}{\mathscr{I}_{rigid}}$
		(b)			(kg m^2)	(keV)	(kg m^2)	
^{174}Yb	70	7.82^5	31.167	0.3081	5.955 × 10^{-54}	76.471^1	2.723 × 10^{-54}	0.4573
^{242}Pu	94	11.90^6	38.834	0.2823	10.184 × 10^{-54}	44.54^2	4.675 × 10^{-54}	0.4591
^{152}Dy	66	17.5^2	28.482	0.7076	6.025 × 10^{-54}	33.75	6.170 × 10^{-54}	1.024

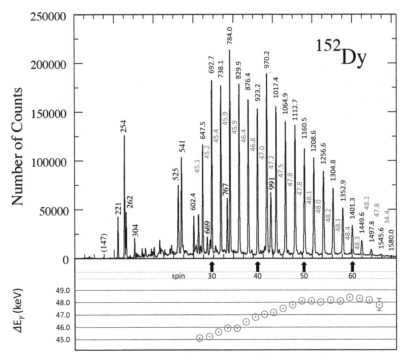

Figure 1.16. Example of a superdeformed band manifested in ^{152}Dy. The figure shows the gamma-ray spectrum associated with the band and the transition energies in keV. Probable spin assignments to the levels from which the gamma rays originate are indicated. The remarkable constancy of the differences between successive gamma rays is depicted at the bottom of the figure. Uncertainties in the gamma-ray energies are generally less than the size of the data points, except the right-hand-most point. Further details are discussed in the text. Note, the de-exciting cascade of gamma rays is attenuated starting around spin 30 due to decay out of the band into lower-lying high-spin states, sometimes called 'draining'; this is beyond the scope of the present discussion. The spectrum was provided courtesy of T Laurtisen and is based on a similar spectrum appearing in [18]. Other data are taken from ENSDF. Reproduced from [23]. Copyright 2010 World Scientific Publishing Company.

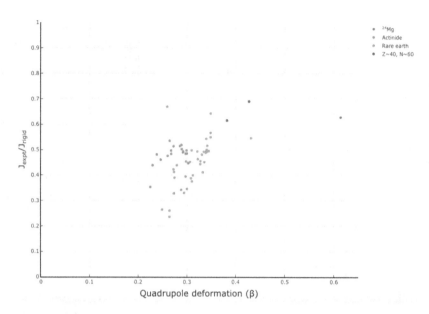

Figure 1.17. Selected view of nuclear moments of inertia for nuclei which exhibit a rotational band built on their ground state. Moments of inertia are presented as a ratio of he experimental moment of inertia deduced from the excitation energy of the 2^+ state (equation (1.1)) with $A = \hbar^2/2\mathscr{J}_{\mathrm{expt}}$, divided by the rigid body moment of inertia (equation (1.24)), plotted as a function of the quadrupole deformation, β for the relevant isotopes. Data are presented for the rare earth isotopes, the actinide region, the region close to $Z \sim 40$, $N \sim 60$ and for ^{24}Mg. Interactive version available in e-book which can be downloaded from http://iopscience.iop.org/book/mono/978-0-7503-5643-5.

some nuclei. We caution that this derivation is based on a naïve view of the nucleus as a quadrupole-deformed constant-density extended object with a sharp surface, i.e. no allowance is made for surface diffuseness or higher multipole deformations of the nucleus. We look at higher multipole deformations later. Figure 1.17 presents moments of inertia, extracted using these simple prescriptions, for selected nuclei. We do not comment on the remarkable similarity of $\mathscr{J}_{\mathrm{expt}}/\mathscr{J}_{\mathrm{rigid}}$ for ^{174}Yb and ^{242}Pu, except to note that this would appear to be independent of the near-identical scaled-energy patterns shown in table 1.2.

There is considerable confusion over moments of inertia extracted from rotational bands in odd-mass nuclei: the origin is in the use of equation (1.1), modified for excitation and spin of a given band head, viz.

$$E = E_0 + AI(I + 1), \tag{1.27}$$

where $I = K, K + 1, K + 2, \ldots$ and the excitation of the band head is given by $E_0 + AK(K + 1)$. It is essential to include the rotation-particle coupling term when making deductions about moments of inertia in odd nuclei. From plots such as depicted in figures 1.11(a)–(d) *et seq.* a useful relationship for replacement of equation (1.27) would take the form

$$E = E_0 + A(I - \alpha)(I - \alpha + 1), \tag{1.28}$$

where α is the horizontal displacement in spin between the odd-mass nucleus and the neighbouring even–even 'core' nucleus. Thus, e.g. in figure 1.13 a, $\alpha = 2.8$ at $I_i = 13/2$. Such a relationship, to our knowledge, has never received attention in the literature; although it has been suggested [19], this was without exploration at the level of presentation of data such as shown herein. It is immediately evident that extracting an 'A' parameter from data will give different values depending on whether equation (1.27) or (1.28) is used. Note that if single energy differences are used one arrives at the relationship

$$\Delta E_{I,I-2} = E_\gamma = A(4I - 2) \tag{1.29}$$

from equation (1.27) and

$$\Delta E_{I,I-2} = E_\gamma = A(4I - 4\alpha - 2) \tag{1.30}$$

from equation (1.28); whereas if double energy differences are used (employing states with spins $I, I - 2$, and $I - 4$) one arrives at the relationship

$$\Delta^2 E = \Delta E_\gamma = 8A. \tag{1.31}$$

This is universally manifested in odd-mass bands and the ground-state bands of neighboring even–even 'core' nuclei, as illustrated in figures 1.11(a)–(d) *et seq.* Commonly, use of equations (1.29) and (1.31) has led to the terminology 'type-I' and 'type-II' moments of inertia, respectively. We leave this issue without further comment.

There is a further puzzling feature manifested in the comparison of rotational bands in odd-mass nuclei with ground-state bands in even-mass nuclei: the unpaired nucleon should exhibit a 'Coriolis effect' whereby the 'alignment' of its spin, j with the core angular momentum, R should progressively increase with increasing total spin, i.e. the semi-classical angle in equation (1.18) should 'close'. This is not observed. The patterns manifested in figures 1.11–1.13, to lowest order, show a constant alignment, i.e., there is no classic rotational alignment effect such as occurs with gyroscopes. Thus, one must question even the idea that the nucleus is 'rotating' in the classical sense when rotation-alignment effects (Coriolis effects) are not conforming to classical patterns.

We note that the above empirical patterns have not been addressed theoretically at a foundational level. Discussion can be found in the literature regarding some of the effects portrayed; but attempts to explain such data have been confined to the naïve rotor model. The result has been to invoke 'delicate cancellations' involving 'blocking of pairing-correlation effects', 'deformation-driving effects', and 'rotation-alignment effects' due to the unpaired nucleon. We express the view that fine-tuning model parameters on a case-by-case basis to fix discrepancies with data is the signature of the much-discussed[7] 'need for a paradigm shift' in science.

---

[7] The expression 'a paradigm shift' was first coined by Thomas S Kuhn in [24]. We add, as authors we did not foresee this paradigm shift, we have only assembled data and noted systematic patterns. We do not offer any theoretical insights beyond recommending that the naïve rotor model, while useful for organizing data and describing $E2$ properties of nuclei, should not be the basis for refining exploration of rotational energies in nuclei. We discuss this in more detail later.

A summary view of the data presented can be made with a few key points. Where data are available, in nuclei that can be described as well-deformed, the simplest rotor model is realized at the level of 1%–2% in electromagnetic properties; in nuclei that can be described as superdeformed, the simplest rotor model is realized at the few percent level with respect to rigid-body rotation with a moment of inertia that matches expectations of classical mechanics. However, the requisite high-precision data are severely lacking. Adopting the view that in complex systems, the emergence of such simple behaviour is not by chance, rather it is a manifestation of asymptotic behaviour, this indicates that much work needs to be done.

1.5 Exercises

1-1. With reference to table 1.1, for the nuclei shown, tabulate values of $B_{I,I-2}/B_{20} f(I)$ for 12 → 10, 14 → 12, ... transitions using $B(E2)$ data in ENSDF. Note: use equation (1.6) to determine $f(12)$, $f(14)$, ...

1-2. Make a similar table to table 1.1 for the actinide isotopes shown in figure 1.2 using ENSDF data for $B(E2)$ values.

1-3. Following on from exercise 1-2, find and add other deformed actinide region $B_{I,I-2}$ data using ENSDF data for $B(E2)$ values. (Note: the data are very limited and confined to low spin states.)

1-4. With reference to table 1.2, using data in ENSDF, explore deformed rare earth and actinide region nuclei for similar scaled rotational transition energies in even–even nuclei. As a set of useful starting nuclei, we suggest: ^{156}Nd, ^{158,160}Sm, ^{240,242}Pu, ^{172,174}Yb, ^{168,170}Er, ^{180}Hf; then look at ^{160}Gd, ^{162,164}Dy, ^{166}Er, ^{170,176}Yb, ^{182}W, ^{234,236,238}U, ^{236,238}Pu.

1-5. Test the rotational energy formula, equation (1.3), for the nuclei presented in table 1.2.

1-6. Using equation (1.12), obtain the Q_0^{gs} values in figures 1.7(a)–(f) from the Q_s^{gs} values given in ENSDF.

1-7. With reference to figure 1.11, for E_γ versus I_i using data in ENSDF, make similar plots for rotational bands built on the Nilsson configuration 7/2+ [404]. As a set of useful starting nuclei, we suggest odd-mass Lu and Ta isotopes where this Nilsson configuration forms the ground-state bands.

1-8. For the nucleus ^{157}Tb, make a plot of E_γ versus I_i, cf those shown in figures 1.11–1.13, for the Nilsson configuration 5/2− [532]. Comment on the statement in the narrative that alignment effects increase with the decreasing value of the Nilsson quantum number Ω, by comparing with figure 1.12 and noting that 5/2− [532], 7/2− [523] and 9/2− [514] all have the spherical parentage configuration $h_{11/2}$, i.e. $j = 11/2$. (Use the data given in ENSDF for ^{157}Tb, where note that the head of the 5/2− [532] band is at 326.3 keV.)

1-9. For the scaling Q_0/ZR^2, compare typical rare earth nuclei (figures 1.8 and 1.9) and actinide nuclei (figure 1.10) Q_0 values. Use $R = 1.2A^{1/3}$ fm. Estimate the Q_0 values by inspection of the figures.

1-10. If j is parallel to I, for an even–even neighbouring 'core' band with $I = 0, 2, 4, 6, \ldots$, sketch what a plot of E_γ versus I_i will look like in an odd-mass nucleus compared to the neighbouring even–even nucleus.

1-11. Derive equation (1.9) from equations (1.4), (1.5), and (1.8). Note, a Clebsch–Gordan coefficient calculator is available on Wolfram Alpha (https://www.wolframalpha.com/input?i=Clebsch-Gordan+calculator).

1-12. Look for superdeformed bands that exhibit near constant differences in transition energies such as illustrated in figure 1.16 for ^{152}Dy. Relevant data can be obtained from [20].

1-13. Explore fitting equation (1.28) to odd-A nuclei. Start with the nuclei in figure 1.13 and use values of α given by the numbers in the black boxes, e.g. $\alpha = 2.2$ for ^{173}Yb. How do the fits compare with using equation (1.27)?

1-14. There is a simple empirical correlation between $B(E2: 0_1^+ \rightarrow 2_1^+)$ values and the excitation energy of the first excited 2^+ state, $E(2_1^+)$, called the Grodzins relationship [21], viz.

$$B(E2: 0_1^+ \rightarrow 2_1^+) \times E(2_1^+) \times \frac{A}{Z^2} \sim \text{constant.} \tag{1.32}$$

This is often employed with the dimensions of $B(E2)$ in e^2b^2 and $E(2)$ in keV, whence the constant for many nuclei is about 16. Note that ENSDF usually gives $B(E2)$ as $B(E2; 2_1^+ \rightarrow 0_1^+)$ in W.u.; recall $B(E2: 0_1^+ \rightarrow 2_1^+) = 5 \times B(E2; 2_1^+ \rightarrow 0_1^+)$, cf equation (1.4) (the spin factor in the denominator). The relationship between $B(E2)$ values in W.u. and e^2b^2 is given by 1 W.u. $= 5.940 \times 10^{-6} A^{4/3}$ e^2b^2. The $B(E2; 0_1^+ \rightarrow 2_1^+)$ values in e^2b^2 are compiled in [22]. $E(2)$ values in keV are given in ENSDF.

For example, for ^{174}Yb: $5.85 \times 76.471 \times 174/70^2 = 15.9$. The ENSDF value for ^{174}Yb $B(E2; 2_1^+ \rightarrow 0_1^+)$ is 201 W.u.: $201 \times 5 \times 5.94 \times 10^{-6} \times 172^{4/3}$ $= 5.71$ e^2b^2. This value is superseded by the value of 5.85 e^2b^2 given in [22].

Explore the Grodzins relationship using data from ENSDF: for example, look at local mass regions to determine the constancy of the product; compare this product for nuclei at closed shells with nuclei in mid-open shell regions.

1-15. Preamble: the following notes and exercises are at a level considerably more advanced than the preceding exercises. For the less specialized reader it is sufficient to treat Clebsch–Gordan coefficients as numbers that can be obtained using the Wolfram Alpha website. We illustrate in the following how to derive algebraic expressions for $E2$ properties presented in this chapter starting from more general equations given in [10]. We recognize that a typical experimentalist will likely start with data and the equations provided, and reach a familiarity of the effectiveness of the equations without initial concern for their origin[8]. Curiosity may then

---

[8] We note that the famous equation, $E = mc^2$ is widely used, but its derivation is a mystery to many.

lead the reader to the origin of the equations below, which in turn have been derived by methods which are presented from first principles, i.e. derivation of the equations given in appendix A, in [10].

The Clebsch–Gordan coefficients that appear in equations (1.5) and (1.7) can be obtained from appendix A, table A.4 of [10] which gives expresssions for $\langle j_1 m_1 2 m_2 | j m \rangle$. We delay some substitution of assigned values so that the origins of the terms can be seen in [10].

For $j = j_1 - 2$, $m_2 = 0$, $m_1 = m$, and

$$\langle j_1 m 2 0 | j m \rangle = \frac{\sqrt{3(j - m + 2)(j - m + 1)(j + m + 2)(j + m + 1)}}{\sqrt{(2j_1 - 2)(2j_1 - 1)j_1(2j_1 + 1)}}, \tag{1.33}$$

then, for $j_1 = I$, $m = 0$, $j = I - 2$ and

$$
\begin{aligned}
\langle I 0 2 0 | I - 2, 0 \rangle &= \frac{\sqrt{3I(I - 1)I(I - 1)}}{\sqrt{2(I - 1)(2I - 1)I(2I + 1)}} \\
&= \frac{\sqrt{3I(I - 1)}}{\sqrt{2(2I - 1)(2I + 1)}}.
\end{aligned}
\tag{1.34}
$$

For $j = j_1$, $m_2 = 0$, $m_1 = m$, and

$$\langle j_1 m 2 0 | j m \rangle = \frac{3m^2 - j(j + 1)}{\sqrt{(2j_1 - 1)j_1(j_1 + 1)(2j_1 + 3)}}, \tag{1.35}$$

then, for $j = j_1 = I$, $m = 0$,

$$
\begin{aligned}
\langle I 0 2 0 | I 0 \rangle &= \frac{-I(I + 1)}{\sqrt{(2I - 1)I(I + 1)(2I + 3)}} \\
&= -\frac{\sqrt{I(I + 1)}}{\sqrt{(2I - 1)(2I + 3)}},
\end{aligned}
\tag{1.36}
$$

and for $j_1 = I$, $m = I$, $j = I$

$$
\begin{aligned}
\langle I I 2 0 | I I \rangle &= \frac{3I^2 - I(I + 1)}{\sqrt{(2I - 1)I(I + 1)(2I + 3)}} \\
&= \frac{I(2I - 1)}{\sqrt{(2I - 1)I(I + 1)(2I + 3)}} \\
&= \frac{\sqrt{I(2I - 1)}}{\sqrt{(I + 1)(2I + 3)}}.
\end{aligned}
\tag{1.37}
$$

(a) From equation (1.34), obtain $\langle 2 0 2 0 | 0 0 \rangle = \frac{1}{\sqrt{5}}$.

(b) From equation (1.34), obtain equation (1.6).

(c) From equations (1.7), (1.36) and (1.37), obtain equation (1.8).

(d) From equation (1.35) with $j = j_1 = I$, $m = K$ and equation (1.37), adapting equation (1.7) by the replacement $\langle I, K=0|T(E2)|I, K=0\rangle \rightarrow \langle I, K|T(E2)|K\rangle$, obtain equation (1.12).

(e) The factor $(5/16\pi)^{1/2}$ that appears in equation (1.5) (and following equations) is because the electric quadrupole moment of a nucleus is defined as

$$\langle Q \rangle = (16\pi/5)^{1/2} \sum_{\text{nucleons},\, i} e_{\text{eff}}^i r_i^2 Y_{20}(\theta_i, \psi_i), \tag{1.38}$$

where (θ_i, ψ_i) are polar coordinates defined with respect to the centre of mass of the nucleus, Y_{20} is a spherical harmonic, and e_{eff} allows for effective charges on the proton and neutron. Show that the factor $(16\pi/5)^{1/2}$ ensures that the total charge in a nucleus is equal to $+Ze$, where Z is the number of protons (here assume $e_{\text{eff}}^{\text{proton}} = +1e$ and $e_{\text{eff}}^{\text{neutron}} = 0$).

A View of Nuclear Data

1. Some preliminary views of nuclei

In anticipation of addressing simple features observed in nuclear data, some simple criteria of physics philosophy are introduced.

Notably, some basic quantum mechanical concepts are sketched.

The use of models for handling complex many-body systems is noted.

Caution is advised with respect to imposing ideas from familiar avenues of physics

Tutorial 1.1 Some preliminary views of nuclei. The video can be downloaded from https://doi.org/10.1088/978-0-7503-5643-5.

A View of Nuclear Data

Nuclear rotation: a first look

2. Electric quadrupole, E2 properties

The most remarkable emergent phenomenon encountered in nuclear structure study
is deformation and rotation

The primary experimental view of nuclei is γ-ray spectroscopy: the γ-ray portion of the electromagnetic spectrum
is dictated by elementary quantum mechanical energy estimates.
Nuclei can be regarded as distributions of electrical charge that act as "antennae" coupled to em fields.

The most extraordinary emergent feature of nuclear rotation is the simplicity of the em coupling strengths.
There are now sufficient data to argue that nuclear charge distributions appear to be (near) rigid as a function of increasing
angular momentum, i.e. nuclei do not behave like liquid drops.

However, simple rotational model estimates of energies do not agree with obser
This is not understood.

The simplest rotor model provides the necessary quantum mechanics to understand basic nuclea

Tutorial 1.2 E2 properties. The video can be downloaded from https://doi.org/10.1088/978-0-7503-5643-5.

A View of Nuclear Data

Nuclear rotation: a first look

3. Excitation energies in even-even nuclei

Electromagnetic matrix elements of deformed nuclei support (near) rigid intrinsic quadrupole moments,
but energies do not support a fixed moment of inertia : this is a fundamental contradiction

However rotational energy patterns of deformed nuclei, even though not matching the I(I+1) spin dependence of a
symmetric top, exhibit smooth variations with spin.

Some nuclei also exhibit energy patterns, which when scaled, are identical to within ~0.1% over wide spin ranges.

Generally, moments of inertia have values that are about ~50% of classical rigid-bod

So-called superdeformed bands possibly have moments of inertia equal to the classica

Tutorial 1.3 Excitation energies in even-even nuclei. The video can be downloaded from https://doi.org/
10.1088/978-0-7503-5643-5.

A View of Nuclear Data

Nuclear rotation: a first look

4. Excitation energies in odd-mass nuclei

Rotations of odd-mass nuclei necessitate a simple model: the rotation-particle coupling model.
This is expressed as

$$I = R + J,$$

where I is the total nuclear spin, R is the rotational angular momentum and J is the total spin of the particle.

The model leads to an $I \cdot J$ energy term which is sometimes called the "Coriolis" term because it resembles the classical analog.

Contributions of this term to rotational energies in odd-mass nuclei are either zero or a constant va

Consequently, there appear to be no Coriolis effects in nuclei

Tutorial 1.4 Excitation energies in odd mass nuclei. The video can be downloaded from https://doi.org/10.1088/978-0-7503-5643-5.

A View of Nuclear Data

Nuclear rotation

5. Further look at excitation energies in odd-mass nuclei

Concepts of "alignment-energy" and "alignment-spin" emerge from data.

This leads to concepts of "deformation-alignment" (Nilsson model) and "rotation-alignment" as competing energy factors.

Tutorial 1.5 Further look at excitation energies in odd-mass nuclei. The video can be downloaded from https://doi.org/10.1088/978-0-7503-5643-5.

A View of Nuclear Data

Comprehensive (complete) spectrosco[
18. Multi-spectroscopy study: ^{152}Sm

^{152}Sm has been the subject of a coordinated program of study using a variety of spectroscopic techniques.

The various spectroscopic types are pictorially illustrated.

Details are limited to visual illustration, i.e. technical details of analysis are omitted.

Key insights into the structure of ^{152}Sm are noted.

[This is a program being pursued by JLW in collaboration with Paul Garrett (Univ. of Guelph); and in its early stages, with W. David Kulp (when he was at Georgia Tech.).]

| P.E. Garrett et al., J. Phys. G: Nucl. Part. Phys. **31** S1855 (2005) |
| W.D. Kulp et al., Phys. Rev. C **76** 034319 (2007) |
| W.D. Kulp et al., Phys. Rev. C **77** 061301(R) (2008) |
| P.E. Garrett et al., Phys. Rev. Lett. **103** 062502 (2009) |

Tutorial 1.6 Comprehensive spectroscopy: Multi-spectroscopy study: ^{152}Sm. The video can be downloaded from https://doi.org/10.1088/978-0-7503-5643-5.

A View of Nuclear Data

Comprehensive (complete) spectroscopy
18. Multi-spectroscopy study: ^{152}Sm

^{152}Sm has been the subject of a coordinated program of study using a variety of spe[

The various spectroscopic types are pictorially illustrated.

Details are limited to visual illustration, i.e. technical details of analysis a[

Key insights into the structure of ^{152}Sm are noted.

[This is a program being pursued by JLW in collaboration with Paul Garrett (Univ. of Guelph); and in its early stages, with W. David Kulp (when he was at Georgia Tech.).]

| P.E. Garrett et al., J. Phys. G: Nucl. Part. Phys. **31** S1855 (2005) |
| W.D. Kulp et al., Phys. Rev. C **76** 034319 (2007) |
| W.D. Kulp et al., Phys. Rev. C **77** 061301(R) (2008) |
| P.E. Garrett et al., Phys. Rev. Lett. **103** 062502 (2009) |

Tutorial 1.6 (Continued.). The video can be downloaded from https://doi.org/10.1088/978-0-7503-5643-5.

A View of Nuclear Data

Comprehensive (complete) spectroscopy
18. Multi-spectroscopy study: ^{152}Sm

^{152}Sm has been the subject of a coordinated program of study using a variety of spectroscopic techniques.

The various spectroscopic types are pictorially illustrated.

Details are limited to visual illustration, i.e. technical details of analysis are omitted.

Key insights into the structure of ^{152}Sm are noted.

[This is a program being pursued by JLW in collaboration with Paul Gar
and in its early stages, with W. David Kulp (when he was at Georg

P.E. Garrett et al., J. Phy
W.D. Kulp et al., Phys. R
W.D. Kulp et al., Phys. Rev. C 77 001301(R) (2008)
P.E. Garrett et al., Phys. Rev. Lett. **103** 062502 (2009)

Tutorial 1.6 (Continued.). The video can be downloaded from https://doi.org/10.1088/978-0-7503-5643-5.

A View of Nuclear Data

Comprehensive (complete) spectroscopy
18. Multi-spectroscopy study: ^{152}Sm

^{152}Sm has been the subject of a coordinated program of study using a variety of spectroscopic techniques.

The various spectroscopic types are pictorially illustrated.

Details are limited to visual illustration, i.e. technical details of analysis are omitted.

Key insights into the structure of ^{152}Sm are noted

[This is a program being pursued by JLW in collaboration with Paul G
and in its early stages, with W. David Kulp (when he was at Geo

P.E. Garrett et al., J. P
W.D. Kulp et al., Phys
W.D. Kulp et al., Phys
P.E. Garrett et al., Phys. Rev. Lett. **103** 062502 (2009)

Tutorial 1.6 (Continued.). The video can be downloaded from https://doi.org/10.1088/978-0-7503-5643-5.

References

[1] Karamatskos E T *et al* 2019 Molecular movie of ultrafast coherent rotational dynamics of OCS *Nat. Commun.* **10** 3364

[2] He Y *et al* 2019 Direct imaging of molecular rotation with high-order-harmonic generation *Phys. Rev.* A **99** 053419

[3] Mizuse K, Fujimoto R, Mizutani N and Ohshima Y 2017 Direct imaging of laser-driven ultrafast molecular rotation *J. Vis. Exp* **120** e54917

[4] Hilborn R C and Yuca C L 1996 Spectroscopic test of the symmetrization postulate for spin-0 nuclei *Phys. Rev. Lett.* **76** 2844

[5] de Angelis M, Gagliardi G, Gianfrani L and Tino G M 1996 Test of the symmetrization postulate for spin-0 particles *Phys. Rev. Lett.* **76** 2840

[6] Modugno G, Ignuscio M and Tino G M 1998 Search for small violations of the symmetrization postulate for spin-0 particles *Phys. Rev. Lett.* **81** 4790

[7] DeMille D, Doyle J M and Sushkov A O 2017 Probing the frontiers of particle physics with tabletop-scale experiments *Science* **357** 990–4

[8] Pastor P C, Galli I, Giusfredi G, Mazzotti D and De Natale P 2015 Testing the validity of bose-einstein statistics in molecules *Phys. Rev.* A **92** 063820

[9] Jenkins D G and Wood J L 2021 *Nuclear Data: A Primer* (Bristol: IOP Publishing) https://iopscience.iop.org/book/mono/978-0-7503-2674-2

[10] Heyde K and Wood J L 2020 *Quantum Mechanics for Nuclear Structure: An Intermediate Level View* (Bristol: IOP Publishing) https://iopscience.iop.org/book/mono/978-0-7503-2171-6

[11] Heyde K and Wood J L 2020 *Quantum Mechanics for Nuclear Structure: A Primer* (Bristol: IOP Publishing) https://iopscience.iop.org/book/mono/978-0-7503-2179-2

[12] Kusakari H, Oshima M, Uchikura A, Sugawara M, Tomotani A, Ichikawa S, Iimura H, Morikawa T, Inamura T and Matsuzaki M 2000 Erratum: Electromagnetic transition probabilities in the natural-parity rotational bands of $^{155, 157}$ Gd *Phys. Rev.* C **63** 029901(E)

[13] Stone N J 2021 Table of nuclear electric quadrupole moments Technical report *IAEA INDC (NDS)-0833*

[14] Powers R J, Boehm F, Hahn A A, Miller J P, Vuilleumier J-L, Wang K-C, Zehnder A, Kunselman A R and Roberson P 1977 An experimental study of $E0$ $E2$ and $E4$ charge moments of ^{161}Dy using muonic atoms *Nucl. Phys.* A **292** 487–505

[15] Zehnder A, Boehm F, Dey W, Engfer R, Walter H K and Vuilleumier J-L 1975 Charge parameters, isotope shifts, quadrupole moments, and nuclear excitation in muonic $^{170-174, 176}$ Yb *Nucl. Phys.* A **254** 315–40

[16] Tanaka Y, Steffen R M, Shera E B, Reuter W, Hoehn M V and Zumbro J D 1984 Measurement and analysis of muonic x rays of $^{176, 177, 178, 179, 180}$ Hf *Phys. Rev.* C **30** 350

[17] Zumbro J D, Naumann R A, Hoehn M V, Reuter W, Shera E B, Bemis C E and Tanaka Y 1986 $E2$ and $E4$ deformations in ^{232}Th and $^{239, 240, 242}$ Pu *Phys. Lett.* B **167** 383

[18] Lauritsen T *et al* 2002 Direct decay from the superdeformed band to the yrast line in ^{152}Dy$_{86}$ *Phys. Rev. Lett.* **88** 042501

[19] Peker L K, Pearlstein S, Rasmussen J O and Hamilton J H 1984 Strength of Coriolis alignment in actinide nuclei *Phys. Rev.* C **29** 271

[20] Singh B, Zywina R and Firestone R B 2002 table of superdeformed nuclear bands and fission isomers: 3rd edn (October 2002) *Nucl. Data Sheets* **97** 241–592

[21] Grodzins L 1962 The uniform behaviour of electric quadrupole transition probabilities from first 2^+ states in even–even nuclei *Phys. Lett.* **2** 88

[22] Pritychenko B, Birch M, Singh B and Horoi M 2016 tables of $E2$ transition probabilities from the first 2^+ states in even-even nuclei *At. Data Nucl. Data tables* **107** 1–139

[23] Rowe D J and Wood J L 2010 *Fundamentals of Nuclear Models: Foundational Models* (Singapore: World Scientific)

[24] Kuhn T S 1962 *The Structure of Scientific Revolutions* (Chicago, IL: University of Chicago Press)

IOP Publishing

Nuclear Data
A collective motion view
David Jenkins and John L Wood

Chapter 2

Do nuclei exhibit asymmetric rotor behaviour?

An asymmetric top model is introduced. It possesses independent electric quadrupole and inertia tensors. This is mandated by energies and E2 properties of nuclei. The β and γ parameters describing an ellipsoidal nuclear shape are introduced. The decomposition of the inertia tensor into its three components is demonstrated. The concept of SO(5) symmetry of an ellipsoid is introduced. The connection to and the breakdown of the Davydov model is detailed. ΔK = 2 mixing and the algebra of angular momentum in a body-fixed frame is introduced, leading to the powerful Mikhailov relationship. Many relationships between E2 matrix elements are explored. The breakdown of the model is illustrated, and a leading cause is suggested (which is realized in chapter 6).

Concepts: asymmetric top model, energy patterns, multiple K bands, inertia tensor, electric quadrupole ($E2$) matrix elements, $E2$ sum rules, Kumar–Cline rotational invariants, P_3 and P_4 invariants, SO(5) symmetry of ellipsoids, Davydov model, $\Delta K = 2$ mixing, Mikhailov relationship.

Learning outcomes: The key data view from this chapter is the plausible applicability of a version of the axially asymmetric rotor model to a significant number of even–even nuclei. This model has independent components of the inertia tensor; and further, these are independent of the electric quadrupole tensor. The independence of these basic model parameters provides a simple explanation of unexpected properties of electric quadrupole matrix elements in some nuclei. Notably, some matrix elements are zero and some exhibit simple summation relationships. A two-band mixing version of the model applied to a well-studied deformed nucleus provides a description of electric quadrupole matrix elements accurate at the few-percent level. For nuclei where extensive data are available, the model describes electric quadrupole matrix elements accurate at the tens-of-percent level.

The simplest departure from the basic axially symmetric rotor in quantum mechanics is the extension to allow for an axially asymmetric shape. This is widely

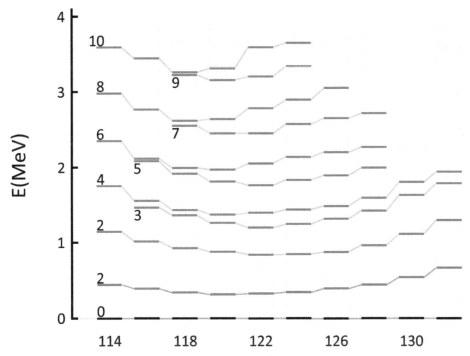

Figure 2.1. Systematics of $K = 2$ bands in the Xe isotopes. States with spin 2, 3, 4, 5, 6, 7, 8, 9 and 10 which are assigned to these bands are shown in blue. The spin-2 members of the ground-state bands are shown for reference, in red. The data are taken from ENSDF.

encountered in molecular rotations, and it appears to be worthy of consideration in the interpretation of nuclear rotations. There is a compelling feature in the excitations observed in even–even nuclei: the second excited states with spin-parity 2^+ occur at low excitation energy in nearly all non-closed shell nuclei. These states exhibit moderately enhanced $B(E2)$ values for their decay to the ground state; thus, they are candidates for so-called 'unfavoured' rotations[1]. Such rotations are a signature of axially asymmetric rotors.

The second excited 2^+ states in deformed nuclei are always observed to exhibit rotational bands. These have the spin sequence $I = 2, 3, 4, 5, 6, \ldots$ and correspond to quanta of favoured rotation in combination with one quantum of unfavoured rotation. An example of an isotopic sequence of such bands is shown in figure 2.1. Indeed, axially asymmetric rotors possess multiple rotational bands, and these can be classified using the K quantum number of the axially symmetric rotor. Thus, the ground-state band can be labelled with '$K = 0$'; the band corresponding to one quantum of unfavoured rotation can be labelled with '$K = 2$'; the band corresponding to two quanta of unfavoured rotation can be labelled with '$K = 4$'; and so on. The four isotopes ^{186,188,190,192}Os are the best examples of this pattern and are shown

---

[1] These states can alternatively be interpreted as vibrational. Such excitations arise as a so-called 'normal mode' of vibration of a spheroidal deformed liquid drop. Details are presented in chapter 6.

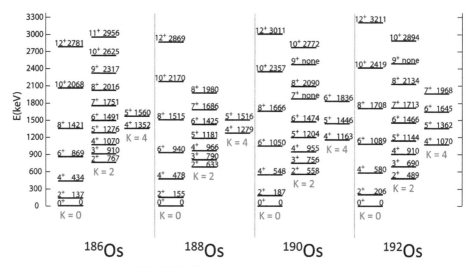

Figure 2.2. Collective bands in $^{186, 188, 190, 192}$Os, classified as having $K = 0$, 2 and 4. Reprinted figure with permission from [1]. Copyright 2008 by the American Physical Society.

in figure 2.2. However, a major note of caution is needed. An axially asymmetric rotor possesses mixing of the K quantum number (and is the reason for the use of quotation marks); this is treated in detail shortly.

2.1 An asymmetric rotor model

To discuss quantized rotations of an axially asymmetric rigid body, we first impose the constraint that the body possesses three orthogonal planes of reflection symmetry passing through its centre of mass. This results in important symmetry constraints on the allowed energy eigenvectors of the system. Thus, we will work in a basis expressed as

$$|IK\rangle = \frac{1}{\sqrt{2}}[|IK\rangle + (-1)^{I+K}|I, -K\rangle], \tag{2.1}$$

even though (as we will see) K is not a good quantum number. The planes of reflection symmetry dictate that the K bands all have even values of K.

The model Hamiltonian[2] that we discuss is expressed as

$$H = A_1 I_1^2 + A_2 I_2^2 + A_3 I_3^2, \tag{2.2}$$

---

[2] This model has been developed in a series of papers [1–5]. It was motivated by failings of the historical approach that goes under the name of the Davydov model [6]. Two key features distinguish this model from the Davydov model: the components of the inertia tensor are independent of parameterization of the $E2$ tensor, and they are independent with respect to each other. The $E2$ properties are expressed at the level of matrix elements, which carry signs that are critical for determining whether the rotor is prolate or oblate. We note that, at the time of the inception of the Davydov model, data for $E2$ matrix elements with their signs were not available; such data are still severely lacking. We provide an extensive compilation of $E2$ matrix elements, which includes many 2_2^+ states for which there are data, in appendix C.

where A_1, A_2, A_3 are model parameters and I_1, I_2, I_3 are the components of angular momentum in the body frame of reference. Note that model parameters are fitted to data and then looked at in terms of components of the inertia tensor. In addition, for the study of electric quadrupole, $E2$ properties of nuclei we must define a model $E2$ operator: we adopt the form

$$T(E2) = \cos\gamma \ T_0^{(2)} + \frac{\sin\gamma}{\sqrt{2}}\left(T_{+2}^{(2)} + T_{-2}^{(2)}\right), \tag{2.3}$$

where the components of the spherical tensor, $E2$ operator $T_\nu^{(2)}$ are reduced by use of the Wigner–Eckart theorem, viz.

$$\langle I_f K_f || T_\nu^{(2)} || I_i K_i \rangle = Q_0(2I_i + 1)\langle I_i K_i; 2, \nu | I_f K_f \rangle \tag{2.4}$$

and thus provide two fitting parameters Q_0 and γ. The parameter Q_0 is already familiar from the symmetric rotor fits to data and is recovered here for $\gamma = 0$; the parameter γ describes the axial asymmetry of the electric quadrupole tensor. In this model, the asymmetry in the inertia tensor resides in the parameters A_1, A_2, A_3 and is defined below.

The Hamiltonian is expressed in the form

$$H = AI^2 + FI_3^2 + G(I_+^2 + I_-^2), \tag{2.5}$$

where

$$A = A_1 + A_2, \quad F = A_3 - A, \quad G = \frac{1}{4}(A_1 - A_2), \tag{2.6}$$

and

$$I_\pm = I_1 \pm iI_2. \tag{2.7}$$

The A and F terms result in energy eigenvalues

$$E = AI(I + 1) + FK^2, \tag{2.8}$$

and the spectrum is depicted in figure 2.3. The G term results in $\Delta K = \pm 2$ mixing: this is evident from the operators I_+^2 and I_-^2, which act as double-lowering and double-raising operators, respectively (note the sign reversal for body-frame operators, see appendix B). The action of the G term is illustrated in the subspace of the two spin-2 configurations, $|\, 2, K = 0\rangle$ and $|\, 2, K = 2\rangle$, shown in figure 2.4. The mixing equations are given by

$$\langle I, K \pm 2 | H | IK \rangle = G\sqrt{(I \mp K)(I \pm K + 1)(I \mp K - 1)(I \pm K + 2)} \tag{2.9}$$

and

$$H(2) = \begin{pmatrix} 6A & 4\sqrt{3}\,G \\ 4\sqrt{3}\,G & 6A + 4F \end{pmatrix}, \tag{2.10}$$

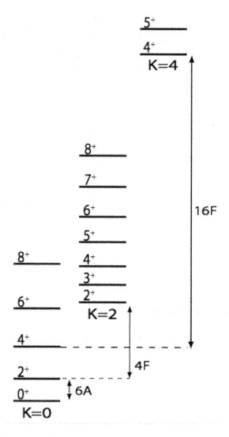

Figure 2.3. Lowest states in the axially asymmetric or triaxial rotor model, organised into bands with $K = 0, 2$ and 4. The energies are defined by equation (2.8).

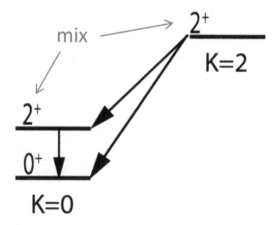

Figure 2.4. The spin-0, 2 subspace of the triaxial rotor model, pointing out the mixing that occurs, cf equations (2.10)–(2.11).

which yields

$$E(2) = 6A + 2F \pm 2\sqrt{(F^2 + 12G^2)}. \tag{2.11}$$

The two equations, embodied in equation (2.11) cannot be solved because they have three unknowns. Extension to the spin-0/spin-2/spin-4 subspace encounters deviations due to the spin-dependence of A and F. Thus, one cannot usefully solve for G. However, as detailed below, for nuclei with sufficient information on $E2$ matrix elements a simple and insightful solution is possible.

The spin dependence of the parameters A and F stems from the failure of simple models to provide a precise description of energies in states classified as rotational bands. The spin dependence of the axially symmetric rotor parameter, A in equation (1.1), is illustrated in 1.3 and table 1.2. The deviations for the axially asymmetric rotor parameter, F in equation (2.8), are severe. For example, with reference to figure 2.2 and using equation (2.8), the energy difference $E(4_3) - E(4_1)$ should equal $4[E(4_2) - E(4_1)]$: the model seriously fails. This has been the reason for a lack of interest in asymmetric rotor models. However, one must consider two factors in assessing the tenability of asymmetric rotor models:

1. Excitation energies expressed as having a rotational origin show features which render the concept of a nuclear moment of inertia questionable as a quantum mechanical version of classical mechanical dynamics. Notably, values of the moment of inertia, as a function of spin, are highly sensitive to the nuclear deformation. Thus, the unfavoured rotations of an axially asymmetric rotor, which are associated with a small deformation, can be anticipated to deviate strongly from the model form manifest in equation (2.8).
2. The excitation energies of the $K = 4$ bands, as manifested in figure 2.2, i.e. >1 MeV, are such that other nuclear degrees of freedom with the same spin, parity and K quantum numbers occur at similar energies. Thus, such $K = 4$ bands may not be pure (unmixed) model states. Indeed, this is spectroscopically established for the osmium isotopes; and it appears to be a general feature of nuclei. Details are given in chapter 6.

By confining the introduction of the model to the spin-0/spin-2 subspace, we establish the basis on which the model can be further tested and how its wider application and breakdown can be understood.

The mixing of the $| 2, K = 2 \rangle$ and $| 2, K = 0 \rangle$ configurations can be expressed as

$$|2_1\rangle = \cos\Gamma|2, K = 0\rangle - \sin\Gamma|2, K = 2\rangle \tag{2.12}$$

and

$$|2_2\rangle = \sin\Gamma|2, K = 0\rangle + \cos\Gamma|2, K = 2\rangle \tag{2.13}$$

where

$$|I, K = 2\rangle = \frac{1}{\sqrt{2}}[|I, 2\rangle + (-1)^I|I, -2\rangle] \tag{2.14}$$

and

$$\tan 2\Gamma = 2\sqrt{3}\,\frac{G}{F}. \tag{2.15}$$

This leads to the relationships

$$\langle 0_1 \| T(E2) \| 2_1 \rangle = \sqrt{5/16\pi}\, Q_0 \cos(\gamma + \Gamma), \tag{2.16}$$

$$\langle 0_1 \| T(E2) \| 2_2 \rangle = \sqrt{5/16\pi}\, Q_0 \sin(\gamma + \Gamma), \tag{2.17}$$

$$\langle 2_1 \| T(E2) \| 2_2 \rangle = \sqrt{25/56\pi}\, Q_0 \sin(\gamma - 2\Gamma), \tag{2.18}$$

$$\langle 2_1 \| T(E2) \| 2_1 \rangle = -\sqrt{25/56\pi}\, Q_0 \cos(\gamma - 2\Gamma) = -\langle 2_2 \| T(E2) \| 2_2 \rangle. \tag{2.19}$$

These relationships are connected to transition strengths via

$$B(E2;\, I_i \to I_f) = \frac{\langle I_f \| T(E2) \| I_i \rangle^2}{(2I_i + 1)} \tag{2.20}$$

and to spectroscopic quadrupole moments via

$$Q(2_1^+) = -2/7 Q_0 \cos(\gamma - 2\Gamma) = -Q(2_2^+). \tag{2.21}$$

The relationships between the matrix elements can be usefully depicted using right-angled triangles, as shown in figures 2.5(a) and (b).

The basic relationships involving the model parameters are given by the seven equations (2.11), (2.16)–(2.19); note, equation (2.11) involves two relationships that connect to $E(2_1)$ and $E(2_2)$ and equation (2.19) involves two relationships that connect to $Q(2_1)$ and $Q(2_2)$, as given in equation (2.21). These relationships are expressed in the five model parameters, A, F, G, Q_0, and γ; note that Γ is related to G by equation (2.15). The parameters A, F, and G are energy parameters and the parameters Q_0 and γ describe the $E2$ properties of nuclei. The replacement of G by Γ enables G to be fitted to $E2$ data. Note that there is a conventional ordering $A_1 < A_2 < A_3$, which results in $G < 0$. This is reflected in the phase convention manifested in equation (2.12), i.e., '$-\sin\Gamma$'. With seven equations in five unknowns, parameter values can be fixed in a variety of ways. This is illustrated in figure 2.6 for ^{186}Os. Generally, available data and relative uncertainties favour the use of the relationships in figure 2.5(a).

The model can be viewed through its spin subspaces. With reference to figure 2.4: there is one state for spin 0, no states with spin 1, two states with spin 2 (as already discussed), one state with spin 3, three states with spin 4, and two states with spin 5. With the continuation of K values, viz. $K = 6, 8, 10, \ldots$ (recall, even values for K are dictated by equation (2.1)), there are four states with spin 6, three states with spin 7, and so on. Thus, the matrix that must be diagonalized for spin 4 is, following equations (2.8) and (2.9),

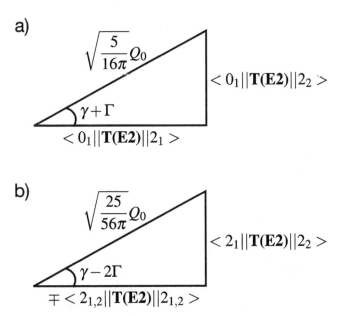

Figure 2.5. (a and b) In (a) the relationship between equations (2.16) and (2.17) is depicted. In (b) the relationship between equations (2.18) and (2.19) is depicted. Reprinted figure with permission from [1]. Copyright 2008 by the American Physical Society.

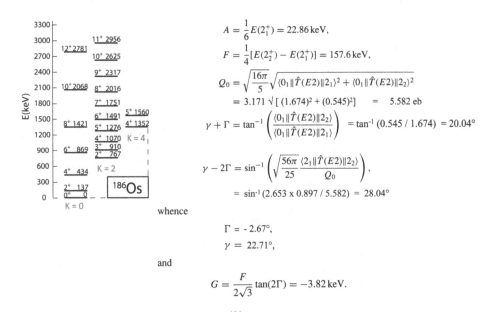

Figure 2.6. The triaxial rotor model applied to ^{186}Os, showing in detail the extraction of the five model parameters A, F, G, Q_0 and γ (note the relationship between Γ and G, cf equation (2.15)). The data are taken from ENSDF.

$$H(4) = \begin{pmatrix} 20A & 12\sqrt{5}\,G & 0 \\ 12\sqrt{5}\,G & 20A + 4F & 4\sqrt{7}\,G \\ 0 & 4\sqrt{7}\,G & 20A + 16F \end{pmatrix} \qquad (2.22)$$

and the matrix which must be diagonalised for spin 6 is

$$H(6) = \begin{pmatrix} 42A & 4\sqrt{210}\,G & 0 & 0 \\ 4\sqrt{210}\,G & 42A + 4F & 6\sqrt{30}\,G & 0 \\ 0 & 6\sqrt{30}\,G & 42A + 16F & 2\sqrt{66}\,G \\ 0 & 0 & 2\sqrt{66}\,G & 42A + 36F \end{pmatrix}. \qquad (2.23)$$

Note that the off-diagonal matrix elements decrease as one moves diagonally 'downwards to the right': thus, for $H(4)$, $12\sqrt{5} = 26.8$, $4\sqrt{7} = 10.6$; and for $H(6)$, $4\sqrt{210} = 58.0$, $6\sqrt{30} = 32.9$, $2\sqrt{66} = 16.2$. Consequently, we can discuss much of the physics by focussing on low-spin states and low-K bands, which is where the mixing dominates. Because detailed data are limited to such a selection of states, there is a natural progression of confrontation between the model and data with increasing spin and K values.

The above basic set of quantum mechanical relationships between the model and observables can be used to explore both higher-spin states and systematics of the parameters across the mass surface. We explore these applications of the model via four separate focal points. First, we look at a global view of the parameter values across the mass surface, which includes a decomposition of the inertia tensor. Second, we look at 'maps' of the empirical $E2$ matrix elements and some deeper relationships that exist between them. Third, we make a generic extension to mixing for all spins in the model for the '$K = 0$' and '$K = 2$' bands. Fourth, we extend the discussion to the full space of the model.

We refer to this model as the general triaxial rotor model, GTRM.

Values of parameters for the GTRM are straightforwardly determined from data as illustrated for ^{186}Os in figure 2.6, and most commonly using the relationships depicted in figure 2.5(a). Of especial interest are the parameters A, F, and G because they can be used to arrive at components of the inertia tensor. We address the parameters Q_0 and γ and make an interpretation of γ versus Γ shortly.

2.2 Moments of inertia of the model

The parameters A, F and G, from their relationship to A_1, A_2 and A_3, cf equation (2.6), and the interpretation of A_1, A_2 and A_3 as inverse moments of inertia, viz.

$$A_1 = \frac{\hbar^2}{2\mathcal{J}_1}, \qquad A_2 = \frac{\hbar^2}{2\mathcal{J}_2}, \qquad A_3 = \frac{\hbar^2}{2\mathcal{J}_3}, \qquad (2.24)$$

lead to

$$\mathcal{J}_1 = \frac{\hbar^2}{2(A + 2G)}, \qquad (2.25)$$

$$\mathscr{I}_2 = \frac{\hbar^2}{2(A - 2G)}, \tag{2.26}$$

$$\mathscr{I}_3 = \frac{\hbar^2}{2(A + F)}. \tag{2.27}$$

From the fit of the $E2$ matrix elements to yield Q_0, γ and Γ, then A, F and G are determined via

$$A = \frac{E(2_1) + E(2_2) - 4F}{12}, \tag{2.28}$$

$$F = \frac{E(2_2) - E(2_1)}{4\sqrt{1 + \tan^2(2\Gamma)}}, \tag{2.29}$$

$$G = \frac{F}{2\sqrt{3}} \tan(2\Gamma). \tag{2.30}$$

Note that the *signs* of the $E2$ matrix elements are critical for obtaining a unique solution to the three components of the inertia tensor; specifically, $\langle 2_1 \|T(E2)\| 2_1 \rangle$ determines whether the electric quadrupole moment is prolate (positive sign) or oblate (negative sign), and the sign of the matrix element product $\langle 0_1 \|T(E2)\| 2_1 \rangle \langle 2_1 \|T(E2)\| 2_2 \rangle \langle 2_2 \|T(E2)\| 0_1 \rangle$ determines whether $\gamma > |\Gamma|$ or $\gamma < |\Gamma|$. We discuss this product quantity in detail shortly.

With respect to the two extremes of rotational dynamics considered for nuclei, rigid rotation and irrotational flow, comparisons of the moments of inertia obtained in equations (2.25)–(2.27), can be made using

$$\mathscr{I}_{\text{rigid},k} = B_{\text{rigid}}\left[1 - 5/4\pi\beta \cos(\gamma - k2\pi/3)\right] \tag{2.31}$$

and

$$\mathscr{I}_{\text{irrot},k} = 4B_{\text{irrot}}\,\beta^2 \sin^2(\gamma - k2\pi/3), \tag{2.32}$$

where $k = 1, 2, 3$,

$$B_{\text{rigid}} = 2/5MR^2 = 0.0138 \times A^{5/3}(\hbar^2/\text{MeV}), \tag{2.33}$$

$$B_{\text{irrot}} = 3/8\pi MR^2 = 0.004\,12 \times A^{5/3}(\hbar^2/\text{MeV}), \tag{2.34}$$

$$\beta = Q_0\frac{\sqrt{5}\,\pi}{3ZR^2}, \tag{2.35}$$

and $R = 1.2A^{1/3}$ fm. Note that there are some subtleties[3] to the mathematical structure of equation (2.32).

Figure 2.7 gives a perspective on the components of the inertia tensor deduced from data. Indeed, experimental data match the mathematical form of equation (2.32). However, it is critical to note that in figure 2.7 all experimental data points are normalized to equation (2.32) for the $k = 1$ component of the inertia tensor. Thus, the absolute values of the experimentally determined inertia components are very different from irrotational values. This pattern reflects that the B_{irrot} parameter has been removed by the normalization. One could interpret this result as reflecting a confirmation of the model origin of equation (2.32).

Although figure 2.7 illustrates a noteworthy match between experimentally determined components of the inertia tensor and equation (2.32), there is a departure as γ approaches 30°. This is more clearly seen in figure 2.8. There is no explanation of this. It would be interesting to find some nuclei that are closer to $\gamma = 30°$, and to determine the model parameters with higher precision. The red line in the figure defines the functional relationship between $\sin(\gamma + \Gamma)$ and γ that exists for the so(5) invariant form manifested in equation (2.32).

2.3 $E2$ matrix element relationships

Figure 2.9 illustrates an important relationship between empirical diagonal quadrupole matrix elements, namely $\langle 2_1 \,||T(E2)||\, 2_1 \rangle = -\langle 2_2 \,||T(E2)||\, 2_2 \rangle$. This is also a property of the GTRM, as evident from equation (2.19). There is no underlying rule stipulating nuclei should possess this property. Some models of nuclear collectivity have this property, and some do not. It can be traced to the reduction of the matrix elements of the model, via the Wigner–Eckart theorem, to a sum rule for Clebsch–Gordan coefficients, i.e., $\langle 2020 \,|\, 20 \rangle + \langle 2220 \,|\, 22 \rangle = 0$. A simple interpretation is that it must be, the lowest two 2^+ states are well 'isolated' from other 2^+ states in these nuclei, i.e. there is no significant mixing with third or fourth 2^+ states. Note

---

[3] The mathematical structure of equation (2.32) is fairly-well manifested in the data; but this mathematical structure is not an indication of irrotational flow. Irrotational flow depends critically on the value of the B parameter and fitted values of B in nuclei do not correspond to irrotational flow. However, the mathematical form of equation (2.32) is the consequence of an so(5) symmetry possessed by the Bohr Hamiltonian. The Bohr Hamiltonian possesses a kinetic energy term which is so(5) invariant; just as a Hamiltonian describing dynamics in the space in which we live, $(3, \mathbb{R})$ possesses a kinetic energy term which is so(3) invariant. The kinetic energy in $(3, \mathbb{R})$ is reducible to a radial kinetic energy and a so-called centrifugal potential. The kinetic energy of the Bohr Hamiltonian, which possesses the geometry of the space $(5, \mathbb{R})$, is reducible to a 'radial' kinetic energy which depends on β and a centrifugal potential which has the functional form given by equation (2.32). This forms an so(5) centrifugal potential.

To take the correspondence between the above-defined centrifugal potentials further: in $(3, \mathbb{R})$ the interest is in potential energy functions which are rotationally symmetric and define so-called central-force problems; in $(5, \mathbb{R})$ the interest is in potential energy functions which range from γ-independent potentials to γ-dependent potentials. A potential which is γ independent possesses so(5) symmetry; a potential with γ dependence breaks so(5) symmetry. An example of a γ-independent potential is the so-called Wilets–Jean potential. An example of a γ-dependent potential is the rigid axially symmetric rotor, where the γ dependence confines the nuclear shape to $\gamma = 0$ with an infinitely deep and infinitely narrow form (hence, the 'freezing' of the γ parameter to the value $\gamma = 0$ removes its appearance from all equations.)

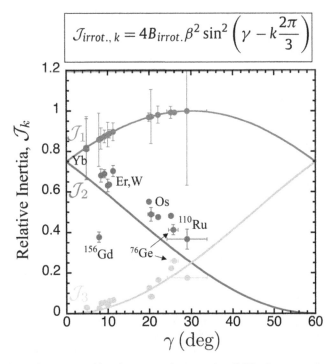

$$\mathcal{J}_{irrot.,k} = 4B_{irrot.}\beta^2 \sin^2\left(\gamma - k\frac{2\pi}{3}\right)$$

Figure 2.7. Experimental moments of inertia compared to equation (2.32), shown as a function of γ. All the experimental values for $\mathcal{J}_1$ are normalized to the model value for $\mathcal{J}_{irrot,1}$, cf equation (2.32). Thus, the experimental values $\mathcal{J}_2$ and $\mathcal{J}_3$ are compared to $\mathcal{J}_{irrot,2}$ and $\mathcal{J}_{irrot,3}$. The $k = 3$-axis is the model symmetry axis @ $\gamma = 0°$ and $60°$. Critical remarks are made in the text regarding the interpretation of this view and its independence of the interpretation of nuclear moments of inertia as due to irrotational flow. The data points for ^{76}Ge and ^{110}Ru are taken from [7] and [8], respectively. The figure has been adapted from one appearing in [5] under CC-BY-4.0 license.

that the K quantum number is at the heart of the relationship. Later we will see that this simple rule does not always work for 4^+ states.

There is a fundamental set of relationships involving the $E2$ operator that goes by the name of the Kumar–Cline sum rules. These relationships involve sums over products of $E2$ matrix elements such that they are rotationally invariant. This abstract statement needs some explanation. For example, the Cartesian components of a vector in three-dimensional space (x, y, z), taken in the product form $x^2 + y^2 + z^2$ is rotationally invariant: it is a scalar, the square of the magnitude of the length of the vector, which is manifestly invariant under rotational transformations. The Kumar–Cline sums are a powerful tool for obtaining rotationally invariant quantities that define the nuclear shape. Here we illustrate a much simpler application of the Kumar–Cline sums: to the matrix elements of the GTRM, as expressed in equations (2.16)–(2.19).

A widely used technique in quantum mechanics for deriving properties of matrix elements and proving mathematical relationships between them is the so-called

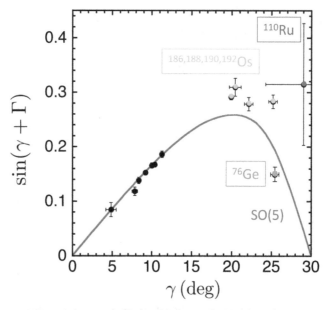

Figure 2.8. Values of sin($\gamma + \Gamma$) versus γ, deduced from experimental data, compared to the relationship fixed between these quantities and labelled 'SO(5)'; this is discussed in the text. The data points for ^{76}Ge and ^{110}Ru are taken from [7] and [8], respectively. See the text for more details. The figure has been adapted from one appearing in [5] under CC-BY-4.0 license.

'resolution of the identity'. This was introduced as a tool already in chapter 4 of [10] as the 'most important tool in the workshop of a quantum mechanic'. In the present context, it provides a way to express relationships between products of $E2$ matrix elements; specifically, the model matrix elements, not the experimentally determined matrix elements. The simplest relationship is for the ground state, viz.

$$\langle 0_1 | T(E2) T(E2) | 0_1 \rangle = \langle 0_1 | T(E2) | 2_1 \rangle \langle 2_1 | T(E2) | 0_1 \rangle$$
$$+ \langle 0_1 | T(E2) | 2_2 \rangle \langle 2_2 | T(E2) | 0_1 \rangle. \tag{2.36}$$

In general, for a nucleus, this sum should be over all spin-2 (positive-parity) states; but in the model space of the GTRM equation (2.36) includes all the spin-2 states in the model. From equations (2.16) and (2.17), viz.

$$\langle 0_1 | T(E2) T(E2) | 0_1 \rangle = 5/16\pi Q_0^2 \cos^2(\gamma + \Gamma) + 5/16\pi Q_0^2 \sin^2(\gamma + \Gamma)$$
$$= 5/16\pi Q_0^2. \tag{2.37}$$

This is a simple algebraic statement of the geometry manifested in figure 2.5(a).

The Kumar–Cline sum for the next 'moment' of the $E2$ operator, a cubic moment, expressed in the model space, is much more revealing than the quadratic moment, viz.

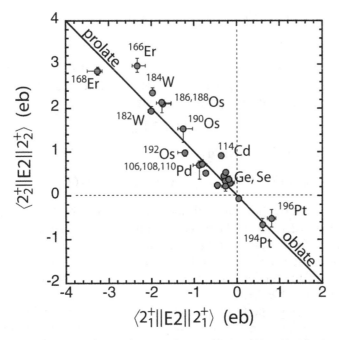

Figure 2.9. Relationship between $\langle 2_2^+ \| E2 \| 2_2^+ \rangle$ and $\langle 2_1^+ \| E2 \| 2_1^+ \rangle$ as manifested in experimental data. (Note $\langle 2_{1,2}^+ \| E2 \| 2_{1,2}^+ \rangle = \langle 2_{1,2}^+ \| T(E2) \| 2_{1,2}^+ \rangle$.) See text for further details. Reprinted figure with permission from [9]. Copyright 2013 by the American Physical Society.

$$
\begin{aligned}
\langle 0_1| T(E2)T(E2)T(E2)|0_1\rangle =\ & \langle 0_1| T(E2)|2_1\rangle\langle 2_1| T(E2)|2_2\rangle\langle 2_2| T(E2)|0_1\rangle \\
& + \langle 0_1| T(E2)|2_2\rangle\langle 2_2| T(E2)|2_1\rangle\langle 2_1| T(E2)|0_1\rangle \\
& + \langle 0_1| T(E2)|2_1\rangle\langle 2_1| T(E2)|2_1\rangle\langle 2_1| T(E2)|0_1\rangle \\
& + \langle 0_1| T(E2)|2_2\rangle\langle 2_2| T(E2)|2_2\rangle\langle 2_2| T(E2)|0_1\rangle,
\end{aligned} \tag{2.38}
$$

whence from equations (2.16)–(2.19) (omitting the numerical factors, which are the same for all these product terms)

$$
\begin{aligned}
\langle 0_1| T(E2)T(E2)T(E2)|0_1\rangle =\ & 2Q_0^3 \cos(\gamma + \Gamma)\sin(\gamma - 2\Gamma)\sin(\gamma + \Gamma) \\
& + Q_0^3 \cos(\gamma + \Gamma)\cos(\gamma - 2\Gamma)\cos(\gamma + \Gamma) \\
& - Q_0^3 \sin(\gamma + \Gamma)\cos(\gamma - 2\Gamma)\sin(\gamma + \Gamma), \\
=\ & Q_0^3 \sin(2\gamma + 2\Gamma)\sin(\gamma - 2\Gamma) \\
& + Q_0^3 \cos(2\gamma + 2\Gamma)\cos(\gamma - 2\Gamma) \\
=\ & Q_0^3 \cos(2\gamma + 2\Gamma + \gamma - 2\Gamma) \\
=\ & Q_0^3 \cos 3\gamma,
\end{aligned} \tag{2.39}
$$

i.e. this moment is independent of Γ, the mixing angle. This reveals that, while quantum mechanical mixing in the spin-2 subspace redistributes the $E2$ strength over the two states, it neither creates nor destroys $E2$ strength.

There is a triple product of $E2$ matrix elements,

$$P_3 = \langle 0_1|T(E2)|2_1\rangle\langle 2_1|T(E2)|2_2\rangle\langle 2_2|T(E2)|0_1\rangle, \tag{2.40}$$

which plays a fundamental role in the interpretation of data. Its sign determines whether the inertia tensor is prolate or oblate. To place this term within a simple picture, we form the quadruple product P_4, viz.

$$P_4 = P_3\langle 2_1|T(E2)|2_1\rangle, \tag{2.41}$$

whence after some algebra, it yields

$$P_4 = 125/7168\pi^2 Q_0^{\,4}\{\cos(4\gamma - 2\Gamma) - \cos 6\Gamma\}. \tag{2.42}$$

Plots of this function (without the scale factor) for $\gamma = 0°$, $5°$, $10°$, $15°$, $20°$, $25°$, $30°$ are shown in figure 2.10. It can be noted that this is the only known collective model that can yield positive values for P_4. There have been a few explorations of terms added to models, specifically to produce $P_4 > 0$; here the possibility emerges as a natural feature of the model. An example of a realization of $P_4 > 0$ in the nucleus ^{194}Pt is depicted in table 2.1.

There is a further property of the model. There is the possibility of destructive interference effects between the triaxiality angle, γ and the mixing angle, Γ. This occurs when $\gamma = 30°$ and $\Gamma = -30°$, whence $\gamma - 2\Gamma = 90°$, with the result that the diagonal matrix elements, cf equation (2.19), $\langle 2_1 | T(E2) | 2_1\rangle$ and $\langle 2_2 | T(E2) | 2_2\rangle$ are zero. This has not been observed so far in any nucleus possessing quadrupole

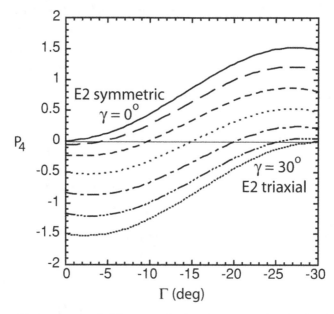

Figure 2.10. Plots of P_4 (see equation (2.42)) versus Γ for the values $\gamma = 0°$, $5°$, $10°$, $15°$, $20°$, $25°$, $30°$. See text for further details. Reprinted figure with permission from [3]. Copyright 2009 by the American Physical Society.

Table 2.1. $E2$ matrix elements for ^{194}Pt compared to the GTRM. The fitted parameter values are $Q_0 = -4.155$ eb, $\gamma = 19.85°$, $\Gamma = -23.92°$. These yield a value for P_4 of $+0.105^7$ which can be compared with an experimental value of $+0.109^{11}$. The experimental data are taken from [11]. The table is taken from [3].

M.E.	Exp. (eb)	Theory (eb)	% dev.
$\langle 0_1 \| T(E2) \| 2_1 \rangle$	$(-)1.281^9$	$(-)1.307^{31}$	-2.0%
$\langle 0_1 \| T(E2) \| 2_2 \rangle$	$(+)0.091^2$	$(+)0.0928^{48}$	2.0%
$\langle 2_1 \| T(E2) \| 2_2 \rangle$	$(-)1.53^5$	$(-)1.449^{34}$	5.1%
$\langle 2_1 \| T(E2) \| 2_1 \rangle$	$+0.61^6$	$+0.595^{16}$	-3.1%
$\langle 2_2 \| T(E2) \| 2_2 \rangle$	-0.66^{14}	-0.595^{16}	9.6%
	Exp. (eb)^4	Theory (eb)^4	% dev.
P_4	$+0.109^{11}$	$+0.105^7$	-4.3%

Table 2.2. $E2$ matrix elements for ^{196}Pt compared to the GTRM. The fitted parameter values are $Q_0 = -3.754$ eb, $\gamma = 20.5°$, $\Gamma = -20.5°$. These yield a value 0.0 for $\langle 0_1^+ \| E2 \| 2_2^+ \rangle$ which can be compared with an experimental value of 0.0. The experimental data are taken from [12]. The table is taken from [4].

M.E.	Exp. (eb)	Theory (eb)	diff (eb)	% diff
$\langle 0_1 \| T(E2) \| 2_1 \rangle$	$(-)1.172^3$	$(-)1.184^{28}$	-0.012	-1.0%
$\langle 0_1 \| T(E2) \| 2_2 \rangle$	0.0	0.0	0.0	0.0%
$\langle 2_1 \| T(E2) \| 2_2 \rangle$	$(-)1.36^5$	$(-)1.243^{51}$	0.117	8.6%
$\langle 2_1 \| T(E2) \| 2_1 \rangle$	$+0.83^9$	$+0.676^{79}$	-0.154	-18.6%
$\langle 2_2 \| T(E2) \| 2_2 \rangle$	-0.51^{21}	-0.676^{79}	-0.166	-32.5%

collectivity. Destructive interference also occurs for $\gamma = -\Gamma$ (not just at $\gamma = 30°$), whence $\gamma + \Gamma = 0°$, with the result that the matrix element, cf equation (2.17), $\langle 0_1 \| T(E2) \| 2_2 \rangle$ is zero. This is observed in the nucleus ^{196}Pt; details are given in table 2.2.

We close this section with the derivation of a simple relationship between $B(E2)$ values and quadrupole moments for the states of spin 2, expressed in terms of the model. From equations (2.16), (2.17) and (2.20), it follows directly that

$$Q(2_1) = -2/7\sqrt{16\pi}\sqrt{B_{20} + B_{2'0} - 7/10B_{2'2}}, \tag{2.43}$$

where $B_{20} = B(E2; 2_1 \rightarrow 0)$, etc and $2' \equiv 2_2$. Equation (2.43) can be compared with equation (1.8) (the scale factor is not simplified, in order to facilitate comparison). Figure 2.11 presents a perspective of selected data viewed using this relationship. Note that it includes the nuclei ^{190,192}Os, which according to figure 1.2 appeared anomalous. The answer is in their triaxial deformation. We address these two nuclei

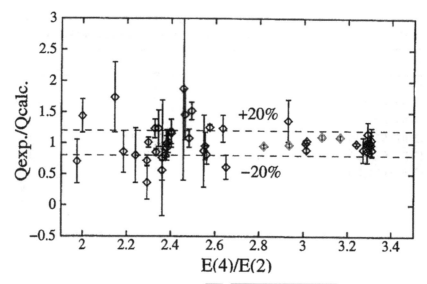

Figure 2.11. Test of equation (2.43), $Q(2_1) = -2/7\sqrt{16\pi}\sqrt{(B_{20} + B_{2'0} - 7/10B_{2'2})}$, expressed as the ratio of experimental value to calculated value. Note that all the strongly deformed nuclei lie at $E(4)/E(2) > 3.2$. The osmium isotopes, [186,188,190,192]Os are highlighted in red. The $B(E2)$ data are taken from ENSDF.

in detail in section 2.5. Further note that, cf equation (2.43), if $7/10B_{2'2} > B_{20} + B_{2'0}$ the model breaks down.

In the details presented above, the energy parameters and the $E2$ operator parameters are all independent. Thus, empirical relationships between them emerge. This feature of the GTRM differs from the historically recognized model of axially asymmetric rotations introduced by A S Davydov. In the Davydov model, the energy parameters are fixed using equation (2.32). Thus, the role of the parameter G in the GTRM would be fixed. The relationship between the parameters Γ and γ is then fixed and is given by

$$\cos 2\Gamma = \frac{\cos 4\gamma + 2\cos 2\gamma}{\sqrt{9 - 8\sin^2 3\gamma}}. \tag{2.44}$$

This relationship would yield a maximum possible value for the GTRM relationship

$$\frac{B_{2'0}}{B_{20}} = \tan^2(\gamma + \Gamma) \tag{2.45}$$

for $\gamma + \Gamma = 15°$, i.e. $B_{2'0}/B_{20} = 0.0718$. Notably, the osmium isotopes, cf figure 2.9, exceed this value. We further note that equation (2.44) yields $\Gamma = -30°$ for $\gamma = 30°$, i.e. $\gamma + \Gamma = 0$ and $\gamma - 2\Gamma = 90°$. The consequences of such 'destructive' interference and its manifestations are discussed above: they appear not to be realized in nuclei for $\gamma = 30°$. Here we note that this destructive interference is inherent in the Davydov model. Details of these relationships are explored in the exercises.

2.4 $\Delta K = 2$ **band mixing**

The GTRM provides a base from which to explore $\Delta K = 2$ band mixing. While this is not an exclusive property of the GTRM, i.e. $\Delta K = 2$ band mixing can be modelled without invoking triaxiality, the model provides a perspective on the origins of such mixing in nuclei. The key to this mixing is in equation (2.5), i.e., the I_+^2 and I_-^2 terms; and their generic matrix elements are given in equation (2.9). One can proceed to work out all the mixed states that occur in a two-band mixing model, notably for a $K = 0$ (ground-state) band and a $K = 2$ (so-called 'gamma') band. From these mixed states, one can then calculate all the $E2$ matrix elements using the operator given in equation (2.3). This is a very tedious process when many states are involved. However, there is an elegant and concise technique that delivers a closed-form algebraic relationship that goes under the name of Mikhailov theory. The approach sketched above involves deriving the energy eigenstates for the Hamiltonian, equation (2.5). In the formal language of quantum mechanics, the $\Delta K = 2$ mixing terms lead to a unitary transformation that delivers a new set of energy eigenstates (a new eigen basis). The elegant approach is to apply the unitary transformation to the $E2$ operator, not the basis states.

Transforming an operator rather than a state basis is not a familiar technique in quantum mechanics. Some perspective on the idea is obtained by studying time-dependent quantum mechanics. Time dependence in quantum mechanical systems is formally developed using a time-evolution operator. This operator provides unitary transformations that yield an eigen basis for all possible times. Two practical representations are introduced: the Schrödinger picture and the Heisenberg picture. In the Schrödinger picture the eigen basis evolves in time; in the Heisenberg picture the operators evolve in time. This is developed in detail in chapter 9 of [10]. (We note that there is a third approach used in time-dependent perturbation theory, called the interaction picture, as developed in chapter 8 of [13]; this is not relevant to the present focus.)

The key to the technique of transforming the $E2$ operator is to introduce the operator which effects the unitary transformation, $\exp(S)$, whence from the expression for a transformation of basis, viz.

$$|\psi'\rangle = \exp(S)|\psi\rangle, \tag{2.46}$$

one can write matrix elements for an operator, e.g. the $E2$ operator, $T(E2)$ in the transformed basis appears as $\langle \psi' \mid T(E2) \mid \psi' \rangle$. Now consider

$$\langle \psi'|T(E2)|\psi'\rangle = \langle \psi|\exp(-S)\ T(E2)\exp(S)|\psi\rangle, \tag{2.47}$$

and instead of regarding this as the action of the operators $\exp(-S)$ and $\exp(S)$ on the bras and kets, it is looked at as the action of these operators on $T(E2)$, viz. $\exp(-S)T(E2)\exp(S)$. Formally, this is a similarity transformation. It enables one to study the action of the operator $T(E2)$ in the unmixed basis $\{|\psi\rangle\}$ by evaluating the expression $\exp(-S)T(E2)\exp(S)$.

To proceed using the technique described above, note that the $E2$ operator is a spherical tensor operator and the operators that are producing the mixing, I_+^2 and I_-^2

have a double-lowering and a double-raising action on states of angular momentum, also on operators with a spherical tensor structure. (We remind the Reader that the raising and lowering operators have reversed signs in the body frame cf appendix A.) Recall, that spherical tensor operators, labelled by λ and μ, possess $2\lambda + 1$ components ($\mu = +\lambda, +\lambda - 1, \ldots, -\lambda$). States of angular momentum, labelled by I, possess $2I + 1$ components. The action of the operator I_+ on a basis state $| IK \rangle$ is given by standard angular momentum algebra as

$$I_+|IK\rangle = \sqrt{(I + K)(I - K + 1)}\big|I, K - 1\rangle \tag{2.48}$$

and for I_-

$$I_-|IK\rangle = \sqrt{(I - K)(I + K + 1)}\big|I, K + 1\rangle. \tag{2.49}$$

These algebraic forms directly carry over (note the formal identities, $I \leftrightarrow \lambda$, $K \leftrightarrow \mu$) to the relationships for a spherical tensor operator (see equation (3.46) in [13]), viz.

$$\left[I_+, T_\mu^{(\lambda)}\right] = \sqrt{(\lambda + \mu)(\lambda - \mu + 1)}\, T_{\mu-1}^{(\lambda)} \tag{2.50}$$

and

$$\left[I_-, T_\mu^{(\lambda)}\right] = \sqrt{(\lambda - \mu)(\lambda + \mu + 1)}\, T_{\mu+1}^{(\lambda)}. \tag{2.51}$$

From the action of the operator I_3 on a basis state $| IK \rangle$, viz.

$$I_3|IK\rangle = K|IK\rangle, \tag{2.52}$$

it follows that

$$\left[I_3, T_\mu^{(\lambda)}\right] = \mu T_\mu^{(\lambda)}. \tag{2.53}$$

The procedure requires an evaluation of the expression on the right-hand side of

$$T(E2)' = \exp(S)\,T(E2)\exp(-S), \tag{2.54}$$

where, cf equation (2.5), with $G/F = G'$

$$S = G'(I_+^2 + I_-^2). \tag{2.55}$$

With the recognition that $T(E2) = T_\mu^{(2)}$, $\mu = +2, +1, 0, -1, -2$, equation (2.54) is 'disentangled' via equations (2.50) and (2.51) using the Baker–Campbell–Hausdorff, BCH, lemma, see below.

We proceed to evaluate equation (2.54) to obtain $T(E2)'$. Using the general BCH form

$$e^{\lambda A}Be^{-\lambda A} = B + \lambda[A, B] + \frac{\lambda^2}{2!}[A, [A, B]] + \frac{\lambda^3}{3!}[A, [A, [A, B]]] + \cdots \tag{2.56}$$

and recognizing that $\lambda(=G')$ is small, so only the λ term, i.e. $[A, B]$ needs to be evaluated, we have

$$T(E2)' = T(E2) + G'[(I_+^2 + I_-^2), T(E2)]. \qquad (2.57)$$

Thus, for example

$$[I_-^2, \ T_0^{(2)}] = I_-[I_-, \ T_0^{(2)}] + [I_-, \ T_0^{(2)}]I_-$$

$$\text{(and dropping the superscript, viz. } T_\mu^{(2)} \to T_\mu) \qquad (2.58)$$

$$= I_-\sqrt{6}\,T_{+1} + \sqrt{6}\,T_{+1}I_-.$$

This can be evaluated further using a derivation from $I^2 = 1/2(I_+I_- + I_-I_+) + I_z^2$, viz.

$$[I^2, \ T_{+2}] = [1/2(I_+I_- + I_-I_+), \ T_{+2}] + \left[I_z^2, \ T_{+2}\right]$$

$$= 1/2(I_+[I_-, \ T_{+2}]I_- + [I_+, \ T_{+2}]I_- + I_-[I_+, \ T_{+2}] + [I_-, \ T_{+2}]I_+) + I_z[I_z, \ T_{+2}] + [I_z, \ T_{+2}]I_z \quad (2.59)$$

$$= 0 + T_{+1}I_- + I_-T_{+1} + 0 + I_z 2T_{+2} + 2T_{+2}I_z$$

$$= T_{+1}I_- + I_-T_{+1} + 2(T_{+2}I_z + I_zT_{+2})$$

whence

$$T_{+1}I_- + I_-T_{+1} = [I^2, \ T_{+2}] - 2(T_{+2}I_z + I_zT_{+2}) \qquad (2.60)$$

and so

$$[I_-^2, \ T_0] = \sqrt{6}[I^2, \ T_{+2}] - 2\sqrt{6}(T_{+2}I_z + I_zT_{+2}); \qquad (2.61)$$

similarly,

$$[I_+^2, \ T_0] = \sqrt{6}[I^2, \ T_{-2}] + 2\sqrt{6}(T_{-2}I_z + I_zT_{-2}). \qquad (2.62)$$

Thus,

$$T(E2)' = T(E2) + \sqrt{6}\,G'\left\{[I^2, \ (T_{+2} + T_{-2})] + 2(T_{-2} - T_{+2})I_z + 2I_z(T_{-2} - T_{+2})\right\}. \quad (2.63)$$

Matrix elements are then evaluated, e.g.,

$$\langle I'K|[I^2, \ T_{-2}]|I, K + 2\rangle = \langle I'K|\left\{I^2 T_{-2} - T_{-2}I^2\right\}|I, K + 2\rangle$$

$$= I'(I' + 1)\langle I'K|T_{-2}|I, K + 2\rangle \qquad (2.64)$$

$$- \langle I'K|T_{-2}|I, K + 2\rangle I(I + 1)$$

$$= [I'(I' + 1) - I(I + 1)]\langle I'K|T_{-2}|I, K + 2\rangle.$$

With straightforward algebra, the general relationship

$$\langle I'2\|T(E2)'\|I0\rangle = \sqrt{2I + 1}\,\langle I022|I'2\rangle(m_0 + m_1[I'(I' + 1) - I(I + 1)]) \qquad (2.65)$$

in lowest order is obtained, where m_0 and m_1 are treated as parameters. Equation (2.65) is called the Mikhailov equation. (We note that many small terms have not been evaluated; many of these terms can be absorbed into the parameters m_0 and m_1. There are higher-order terms, and a further view of these is noted below.)

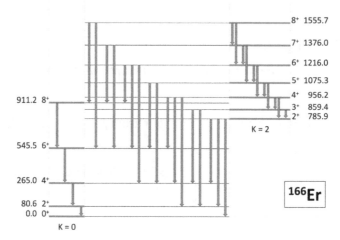

Figure 2.12. Details of the ground-state and gamma bands in ^{166}Er relevant to an analysis of $B(E2)$ values using Mikhailov theory, cf equation (2.65). Excitation energies are given in keV. The data are taken from ENSDF.

Equation (2.65), in the form

$$\frac{\sqrt{B(E2;\ I_i,\ K=2 \to I_f,\ K=0)}}{\langle I_i 22,\ -2|I_f 0\rangle} = m_0 + m_1[(I_f(I_f+1) - I_i(I_i+1)], \qquad (2.66)$$

can be used to make plots of $K=2 \to K=0$ inter-band transition intensities for deformed nuclei. The transitions depicted in figure 2.12 for ^{166}Er are shown in terms of the relationship, equation (2.66) in figure 2.13. Note that the present focus on two-band mixing and equation (2.66) can be extended to higher-order terms in Δ, where

$$\Delta := [I_f(I_f+1) - I_i(I_i+1)]. \qquad (2.67)$$

Thus, noting that the data points in figure 2.13 exhibit an 'S' shaped systematic trend, a fit to

$$\frac{\sqrt{B(E2;\ I_i,\ K=2 \to I_f,\ K=0)}}{\langle I_i 22,\ -2|I_f 0\rangle} = m_0 + m_1\Delta + m_2\Delta^2 + m_3\Delta^3 \qquad (2.68)$$

can be made.

2.5 Breakdown of the model

Identifying where the asymmetric rotor model, GTRM breaks down is critical. We make the point here, and in other places, that we only learn how to advance our understanding via models by identifying their failures.

The model has been applied to a very detailed set of $E2$ matrix elements for ^{186,188,190,192}Os [1]. Table 2.3 presents the comparison between the model and the

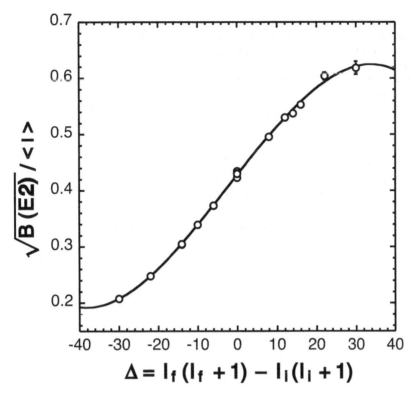

$$\Delta = I_f(I_f + 1) - I_i(I_i + 1)$$

Figure 2.13. A Mikhailov plot, cf equation (2.68), for the gamma-band to ground-band transitions in ^{166}Er. The solid line is a least-squares fit of a cubic polynomial in Δ. Experimental data are given as open circles. Error bars are generally smaller than these circles. Note that there are two data points at $\Delta = -14$, two data points at $\Delta = -6$ and four data points at $\Delta = 0$. Reprinted figure with permission from [14]. Copyright 2006 by the American Physical Society.

experimental $E2$ matrix element data [11]. The most extreme failures are highlighted in red. Note that these failures are confined to $E2$ matrix elements with the smallest values. Further failures occur for $E2$ matrix elements involving the 4_3^+ states (not shown here). These failures are caused by mixing with hexadecapole degrees of freedom. Details are presented in chapter 6.

2.6 Exercises

2-1. Make a systematics plot, similar to figure 2.1, for the even-mass $N = 70$ isotones.

2-2. Make a plot of $E(2_1^+)$ versus A for the even–even nuclei in the region $Z > 50$, $N < 82$. In the plot, connect the isotopic sequences and the isotonic sequences by lines.

2-3. With reference to figure 2.2, other mass regions are known where candidate $K = 4$ bands are observed which may be associated with triaxial rotations. For nuclei in the region of ^{108}Mo explore what is known and make figures similar to figure 2.2 showing $K = 0, 2, 4$ bands.

Table 2.3. Description of $E2$ matrix elements in ^{186,188,190,192}Os using the GTRM. The quantities tabulated are the calculated values of the matrix elements in eb and the percentage deviation from the experimental values, which are not shown but are given in [11]. The focus is on the degree to which the model fails. The largest failures, i.e. >70%, are highlighted as red text. Note that the extreme failures (>300%) involve the smallest matrix elements. Where no percentage deviation is given, there are no useful data for comparison. Further details are given in [1].

	^{186}Os	^{188}Os	^{190}Os	^{192}Os
$2_1 \to 0_1$	1.6697(−0.3%)	1.5812(−0.2%)	1.5190(−0.7%)	1.4414(−1.0%)
$4_1 \to 2_1$	2.7170(−1.6%)	2.5761(−2.5%)	2.4583(+5.0%)	2.3628(+11.7%)
$6_1 \to 4_1$	3.512(−9.7%)	3.338(+0.8%)	3.2369(+9.0%)	3.1448(+7.3%)
$8_1 \to 6_1$	4.237(−1.9%)	4.035(+1.6%)	4.009(+7.8%)	3.840(+7.3%)
$4_2 \to 2_2$	1.7509(−10.9%)	1.661(−6.7%)	1.6115(−13.9%)	1.5580(−4.8%)
$6_2 \to 4_2$	2.865(+3.0%)	2.668(+8.5%)	2.224(−14.5%)	2.172(+3.9%)
$8_2 \to 6_2$	3.550(+8.9%)	3.303(+29.5%)	3.105(+19.4%)	2.906(+25.8%)
$2_2 \to 0_1$	0.5581(+2.4%)	0.4958(+2.6%)	0.4800(+8.1%)	0.4771(+11.0%)
$2_2 \to 2_1$	0.8668(−3.4%)	0.8362(−3.3%)	0.9888(−7.2%)	1.1406(−7.3%)
$2_2 \to 4_1$	0.2949(+29.9%)	0.3072(−18.7%)	0.401(+111.0%)	0.455(+30.0%)
$4_2 \to 2_1$	0.3471(−17.2%)	0.2357(−16.7%)	0.0572(−71.8%)	−0.0402(−130.9%)
$4_2 \to 4_1$	1.2524(+2.7%)	1.187(+7.9%)	1.2849(−10.5%)	1.309(−3.1%)
$4_2 \to 6_1$	0.634(−5.4%)	0.640(+12.2%)	0.867(+31.3%)	0.587(+46.8%)
$6_2 \to 4_1$	0.1535(−52.8%)	0.0141(−88.9%)	−0.3927(−301.4%)	−0.1797(−360.4%)
$6_2 \to 6_1$	1.406(+2.6%)	1.276(−12.6%)	1.123(−36.2%)	1.105(−25.9%)
$2_1 \to 2_1$	−1.917(−9.6%)	−1.795(−3.8%)	−1.627(−30.2%)	−1.411(−16.7%)
$4_1 \to 4_1$	−2.218(−9.8%)	−2.017(−0.8%)	−1.576(−23.1%)	−1.104(−51.3%)
$6_1 \to 6_1$	−2.261(−35.4%)	−1.987(−24.2%)	−1.170(−28.6%)	−0.822(+29.2%)
$8_1 \to 8_1$	−2.160(+4.4%)	−1.874(−35.8%)	−1.234(−31.3%)	−0.719(+45.1%)
$2_2 \to 2_2$	1.917(−9.6%)	1.795(−14.5%)	1.627(+6.3%)	1.4115(+43.3%)
$4_2 \to 4_2$	−1.179(−5.3%)	−1.136(+6.9%)	−1.102(+15.6%)	−0.826(+0.5%)
$6_2 \to 6_2$	−2.168(∅)	−1.938(−45.7%)	−0.818(−2.2%)	−0.751(+44.4%)
$8_2 \to 8_2$	−2.547(∅)	−2.181(∅)	−1.484(−41.3%)	−0.999(−9.7%)

2-4. With reference to figure 2.6, using data in appendix C, make similar 'worksheets' for ^{188,190,192}Os.

2-5. Using ENSDF, identify all mass regions with 2_2^+ states that have energies below 800 keV.

2-6. From equations (2.3), (2.12)–(2.14), making insertion of these expressions into the left-hand side of equation (2.16), show that the expression on the right-hand side of equation (2.16) is obtained.

2-7. For the mixing of the $|2, K = 0\rangle$ and $|2, K = 2\rangle$ configurations, as expressed in equation (2.10), it is convenient to use the condensed notation for two-state mixing introduced in chapter 6 of [10], equations (6.52)–(6.60). Thus, expressing H(2) as

$$\begin{pmatrix} \varepsilon_1 & \nu \\ \nu & \varepsilon_2 \end{pmatrix} = \begin{pmatrix} 6A & 4\sqrt{3}\,G \\ 4\sqrt{3}\,G & 6A + 4F \end{pmatrix} \tag{2.69}$$

and defining

$$\varepsilon_{av} = 1/2(\varepsilon_1 + \varepsilon_2), \tag{2.70}$$

$$\varepsilon = 1/2(\varepsilon_1 - \varepsilon_2) \tag{2.71}$$

i.e.

$$\varepsilon_{av} = 6A + 2F, \tag{2.72}$$

$$\varepsilon = -2F, \tag{2.73}$$

$$\nu = 4\sqrt{3}\,G, \tag{2.74}$$

then equation (6.60) in [10] yields

$$\lambda = \varepsilon_{av} \pm \sqrt{\varepsilon^2 + \nu^2}$$
$$= 6A + 2F \pm \sqrt{4F^2 + 48G^2}, \tag{2.75}$$

from which equation (2.11) follows.

The condensed notation, ε_{av}, ε, ν, helps to keep track of the mixing amplitudes $\{\alpha_1, \beta_1\}$ and $\{\alpha_2, \beta_2\}$ of the eigenvectors associated with the eigenvalues λ_1, λ_2, respectively. Thus, from equations[4] 6.64–6.71 in [10] with

$$\beta_1 = (\lambda_1 - \varepsilon_1)\alpha_1/\nu \tag{2.76}$$

and defining

$$\Delta = (\varepsilon_1 - \lambda_1)/\nu, \tag{2.77}$$

on normalization we get

$$\alpha_1 = \frac{1}{\sqrt{1 + \Delta^2}}. \tag{2.78}$$

So, we obtain

$$\Delta = \frac{-2F + 2\sqrt{F^2 + 12G^2}}{4\sqrt{3}\,G} \tag{2.79}$$

---

[4] Note, there is a typographical error in equation (6.65) of [10], which should read: $\begin{pmatrix} \alpha_1 \\ \beta_1 \end{pmatrix} = \alpha_1 \begin{pmatrix} 1 \\ -\Delta \end{pmatrix}$.

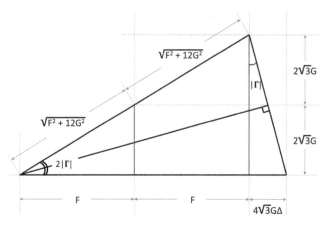

Figure 2.14. Geometrical construction illustrating equation (2.79) and a way to obtain equation (2.80).

and this expression is conveniently viewed geometrically, as shown in figure 2.14. By inspection,

$$\tan 2\Gamma = 2\sqrt{3}\,G/F, \tag{2.80}$$

where we reiterate that $G < 0$ is by convention (because $A_1 < A_2$, i.e. A_1 labels the 'favoured' rotations and so it corresponds to the largest component of the rotational inertia tensor, cf equation (2.24)) and figure 2.14 depicts the magnitude of Γ.

(a) Identify the triangle in figure 2.14 that is associated with equation (2.78), thus connecting α_1 to $\cos\Gamma$, cf equation (2.12).

2-8. The Mikhailov plot in figure 2.13, for ^{166}Er, is generated using primary data from two sources: the absolute $E2$ strength for the transition from the 2_2^+ state to the 0_1^+ (ground) state, the relative $E2$ strengths for the inter-band transitions shown in figure 2.12. From ENSDF, using $B(E2: 2_2^+ \rightarrow 0_1^+) = 5.1721$ W.u. and 1 $e^2b^2 = 184.5$ W.u. [1 W.u. $= 5.940 \times 10^{-6} A^{4/3} e^2 b^2$], we obtain $M_{2'0} = 0.3749$ eb, where $2':=2_2^+$. From ENSDF, the relative intensities and the mixing ratios, $\delta(E2/M1)$ for the transitions shown in figure 2.12 are given in table 2.4.

The data needed for the construction of figure 2.13 are, from left to right, by column:

1. the initial and final spins of the transitions, where, e.g. 8_γ is the 8^+ member of the $K = 2$ or gamma band and 8_g is the 8^+ member of the $K = 0$ or ground band;
2. the gamma-ray transition energy in keV, which can be deduced from differences in level energies given in figure 2.11;
3. relative intensities of gamma rays de-exciting each level, normalized to 100.0 for the strongest gamma-ray;
4. the mixing ratio for transitions with $\Delta I = 0$ or 1 which can occur by $E2$ or M1 decay, denoted by δ, where the $E2$ fraction is given by $\delta^2/(1 + \delta^2)$;

Table 2.4. Input data for making a Mikhailov plot for the gamma band to ground band transitions in ^{166}Er. The details are handled in the text.

Transition	Energy (keV)	Intensity rel.	δ	$B(E2)$ (W.u.)	$\sqrt{B(E2)}$ eb	$\langle \mid \rangle$	$\dfrac{\sqrt{B(E2)}}{\langle \mid \rangle}$
$8_\gamma \rightarrow 8_g$	644.6	86.9^{27}	$+4.9^{23}_{11}$	8.5^9	0.215^{11}	0.6070	0.354^{18}
$8_\gamma \rightarrow 6_g$	1010.3	48.3^6		0.52^5	0.0531^{26}	0.2970	0.179^9
$8_\gamma \rightarrow 6_\gamma$	339.8	100.0^{10}		250^{23}			
$7_\gamma \rightarrow 8_g$	464.8	12.8^6	-63^{12}_{19}	8.0^{16}	0.208^{21}	0.4472	0.465^{47}
$7_\gamma \rightarrow 6_g$	830.6	100.0^{17}	-16.6^{15}_{18}	3.4^7	0.136^{14}	0.5477	0.248^{26}
$7_\gamma \rightarrow 5_\gamma$	300.8	39.16^{23}		220^{40}			
$6_\gamma \rightarrow 8_g$	304.9	0.36^5		1.9^3	0.101^8	0.1961	0.515^{41}
$6_\gamma \rightarrow 6_g$	670.5	100.0^{17}	10.0^{16}_{12}	9.9^7	0.232^8	0.6030	0.385^{13}
$6_\gamma \rightarrow 4_g$	951.0	50.40^{24}		0.88^6	0.0691^{23}	0.3129	0.221^7
$6_\gamma \rightarrow 4_\gamma$	259.7	19.60^{11}		225^{16}			
$5_\gamma \rightarrow 6_g$	529.8	16.63^{27}	-25^4_5	12.4^{15}	0.259^{16}	0.4264	0.607^{38}
$5_\gamma \rightarrow 4_g$	810.3	100.0^{19}	-21.2^{18}_{21}	8.9^{11}	0.220^{14}	0.5641	0.390^{25}
$5_\gamma \rightarrow 3_\gamma$	215.9	4.52^1		300^{40}			
$5_\gamma \rightarrow 4_\gamma$	119.0	0.298^6	1.94^{23}_{21}	310^{40}			
$4_\gamma \rightarrow 6_g$	410.8	1.25^4		2.01^{14}	0.104^4	0.1741	0.597^{23}
$4_\gamma \rightarrow 4_g$	691.3	100.0^6	-3.7^5	11.1^7	0.245^8	0.5922	0.414^{14}
$4_\gamma \rightarrow 2_g$	875.7	54.2^4		1.98^{12}	0.104^3	0.3450	0.300^9
$4_\gamma \rightarrow 2_\gamma$	170.3	1.05^3		138^9			
$3_\gamma \rightarrow 4_g$	594.4	18.82^{17}	-12^2	4.8^9	0.161^{15}	0.3780	0.427^{40}
$3_\gamma \rightarrow 2_g$	778.8	100.0^{24}	-20^2_4	6.6^{12}	0.189^{17}	0.5976	0.316^{28}
$2_\gamma \rightarrow 4_g$	521.0	1.72^4		0.78^4	0.0650^{16}	0.1195	0.544^{13}
$2_\gamma \rightarrow 2_g$	705.3	100.0^{21}	-5^3_{14}	9.6^6	0.228^7	0.5345	0.427^{13}
$2_\gamma \rightarrow 0_g$	785.9	88.9^{18}		5.17^{21}	0.1674^{33}	0.4472	0.374^7

5. the $B(E2)$ value in Weisskopf units (W.u.);

6. $\sqrt{B(E2)}$ in eb, where the conversion from W.u. to eb is given by 1 W.u. $= 5.940 \times 10^{-6} \times A^{4/3}$ e^{2}b^2;

7. the Clebsch–Gordan coefficient for the transition, obtained from Wolfram Alpha, where, e.g. $\langle 822, -2 \mid 80 \rangle = 0.6070$;

8. the plotted quantity, versus $\Delta = I_f(I_f + 1) - I_i(I_i + 1)$, which is, e.g. -30 for the $8_\gamma \rightarrow 6_g$ transition (i—initial, f—final).

Uncertainties are given as, e.g. $86.9^{27} = 86.9 \pm 2.7$. Note that uncertainties quoted for $\sqrt{B(E2)}$ are one half those quoted for $B(E2)$ (as fractions) because of the square-root operation. The data are taken from ENSDF, and see text.

$B(E2)$ values for inter-band transitions such as shown in figure 2.12 are generally of inadequate precision (~10%) for a quantitative test of Mikhailov theory. Thus, relative intensities of inter-band γ-ray transitions, which can be measured to ~1% precision, are used with an absolute normalization procedure which we describe below.

Mikhailov plots, such as shown in figure 2.13, possess two basic features: a slope and an intercept. The slope would be zero, i.e. a horizontal line, if there was no mixing of the two bands. The intercept is defined for the $\Delta I = 0$ inter-band transitions, i.e. $2' \rightarrow 2$, $4' \rightarrow 4$, $6' \rightarrow 6$ and $8' \rightarrow 8$ transitions. The importance of the $2' \rightarrow 0$ transition is that it usually has the most precisely determined $B(E2)$ value, and from the relative γ-ray transition intensities for the $2' \rightarrow 2/2' \rightarrow 0$ transitions (and the $2' \rightarrow 2$ $E2/M1$ mixing ratio) this fixes the intercept of the plot. Thus, for ^{166}Er, from $B(E2; 2' \rightarrow 0) = 5.1721$ W.u. and $I_\gamma(2' \rightarrow 2)/I_\gamma(2' \rightarrow 0) = 100.0/88.9$, recalling that $I_\gamma \sim B(E2) \times E_\gamma^5$, and that the $2' \rightarrow 2$ transition has an M1 admixture and so a factor $\delta^2/(1 + \delta^2)$ must be applied, from the data in table 2.4,

$$B(E2; 2' \rightarrow 2) = 5.17 \times 100.0/88.9 \times (785.90/705.33)^5 \times (5^2/(1 + 5^2))$$
$$= 9.60 \text{ W.u.} \tag{2.81}$$

This is equivalent to $9.60/184.5 = 0.052\,05$ e^{2}b^2. The value to be plotted for the Mikhailov relationship is $\sqrt{B(E2)}/\langle 222, -2 \,|\, 20 \rangle$, where the denominator is a Clebsch–Gordan coefficient and has the value 0.5345, whence the intercept is fixed at 0.4268.

The slope is fixed from the values for $\sqrt{B(E2; 2' \rightarrow 0)}/\langle 222, -2 \,|\, 00 \rangle$, $\sqrt{B(E2; 2' \rightarrow 2)}/\langle 222, -2 \,|\, 20 \rangle$, and $\sqrt{B(E2; 2' \rightarrow 4)}/\langle 222, -2 \,|\, 40 \rangle$,

(a) Obtain the value for $\sqrt{B(E2; 2' \rightarrow 4)}/\langle 222, -2 \,|\, 40 \rangle$ using the relative γ-ray intensity given in table 2.4 and confirm with the Clebsch–Gordan coefficient calculator available at Wolfram Alpha https://www.wolframalpha.com/input?i=Clebsch-Gordan+calculator.

Two values for the slope are provided: from the plot points $(-6, 0.3743)$ and $(0, 0.4268)$, cf equation (2.68) (linear form, i.e. ignore quadratic and cubic terms), $m_1 = 0.008\,75$ e^{2}b^2; and from the plot points $(0, 0.4268)$ and $(+14, [0.544])$, $m_1 = [0.008\,36\ \text{e}^2\text{b}^2]$. The average value $0.008\,55$ e^{2}b^2 can be adopted.

For precision tests, branching ratios for other states are then used, with the normalization of one of the branches to the building up of the Mikhailov plot. Thus, for the spin-3 state, the points $(-6, 0.316)$ and $(+8, 0.427)$ are renormalized by $0.4316 \rightarrow 0.427 - 6 \times 0.008\,55 = 0.376$, i.e. $0.427 \rightarrow 0.427 \times 0.376/0.316 = 0.508$.

2-9. The normalization technique presented in exercise 2-8 is adopted to achieve a high precision test of the Mikhailov relationship. The technique uses relative intensities of decay branches which can be determined to a precision an order of magnitude higher than half-lives and deduced absolute $B(E2)$ values given in ENSDF. In ^{166}Er, there are six half-life values that have been 'bypassed', using this approach, viz. 3_γ $(4.5^8$ ps), 4_γ $(3.5^2$ ps), 5_γ $(2.7^3$ ps), 6_γ $(4.4^3$ ps), 7_γ $(4.9^9$ ps), 8_γ $(3.7^3$ ps). Note, the half-life value for the 2_γ state, 3.12^{10} ps, is used in exercise 2-8 to obtain the intercept of the Mikhailov plot; the values 9_γ $(2.4^5$ ps), 10_γ $(1.78^{17}$ps) are not used herein because their precision is too low.

The renormalization for the 6_γ state, used to make the Mikhailov plot shown in figure 2.13, is presented in figure 2.15. It requires that the half-life for the 6_γ state be changed from 4.4 ps to 3.58 ps, which exceeds the quoted uncertainty (4.4 ± 0.3 ps). We note that the half-life for the 6_γ state, as documented in ENSDF, is based on an unweighted average of half-life values 3.5^4 ps, 4.4^5 ps and 4.6^5 ps. Further note that the branching ratios adopted by ENSDF for the de-excitation of the 6_γ state are $6_\gamma \rightarrow 4_\gamma$ (50.40^{24}), $6_\gamma \rightarrow 6_g$ (100.0^{17}) and $6_\gamma \rightarrow 8_g$ (0.36^5), whereas figure 2.13 is based on the values $6_\gamma \rightarrow 4_g$ (50.26^{16}), $6_\gamma \rightarrow 6_g$ (100.00^{53}) and $6_\gamma \rightarrow 8_g$ (0.416^{15}) taken from [14].

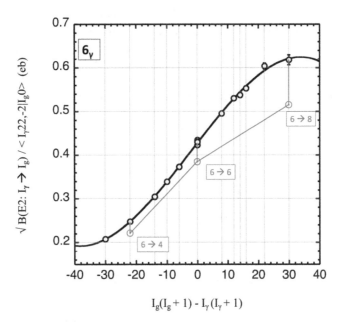

Figure 2.15. Mikhailov plot for ^{166}Er showing the fit presented in figure 2.13 and the results of adopting the half-life for the 6_γ state, 4.4^3 ps given in ENSDF. The half-life deduced from the fit is 3.58 ps and the adoption of this value results in the shifts depicted by the upwards-pointing vertical red arrows. The shifts are not equal in magnitude because of the square-root factor in the plotted quantities.

Show that the Mikhailov plot in figure 2.13 incurs the following renormalized lifetimes:

(a) For the 3_γ state: 4.5 ps → 3.28 ps.
(b) For the 5_γ state: 2.7 ps → 3.42 ps.

2-10. Following the procedures in exercise 2-8, using data in ENSDF, make a Mikhailov analysis for the gamma band in ^{168}Er.

2-11. From exercises 2-8 and 2-9, establish criteria for carrying out systematic Mikhailov analyses of gamma bands in rare earth region nuclei, i.e. what data must be available to initiate such a program?

2-12. Mikhailov analyses of $K = 2 \rightarrow K = 0$ inter-band transitions plot the quantities $B(E2; I_\gamma \rightarrow I_g)/\langle I_\gamma 22, -2 \mid I_g 0\rangle$. There is a similar analysis applicable to $K = 0 \rightarrow K = 0$ inter-band transitions which plots the quantities $B(E2; I_\beta \rightarrow I_g)/\langle I_\beta 020 \mid I_g 0\rangle$. There are major differences between, e.g. $\langle 222, -2 \mid 40\rangle$ and $\langle 2020 \mid 40\rangle$, and therefore between the branching ratios $2_\gamma \rightarrow 4_g/2_\gamma \rightarrow 2_g$ and $2_\beta \rightarrow 4_g/2_\beta \rightarrow 2_g$. By inspection of data in ENSDF for branching ratios from second, third, $\cdots$ excited 2^+ states in rare earth nuclei, suggest assignment of K quantum numbers to these 2^+ states:

(a) ^{166}Er 2^+ 785.9 keV, 2^+ 1528.4 keV;
(b) ^{168}Er 2^+ 821.2 keV, 2^+ 1276.3 keV, 1493.1 keV, 1848.4 keV.

The ratios of Clebsch–Gordan coefficients, as encountered in such branching ratios, are called Alaga rules [15]. Comment on the possibility of using such branching ratios to assign K quantum number values to 2^+ states.

(c) Carry out an analysis similar to the above for 4^+ states.

A View of Nuclear Data

Are nuclei axially asymmetric?
19. A basic model

A model view of axial asymmetry of nuclei is presented.

Axial asymmetry possesses inherent quantum mechanical mixing.

Tutorial 2.1 Are nuclei axially symmetric? A basic model. The video can be downloaded from https://doi.org/10.1088/978-0-7503-5643-5.

A View of Nuclear Data

Are nuclei axially asymmetric?
20. Case study: Os isotopes

Mixing, as described by the simple model-based view, leads to important E2 collective effects which are matched by data.

The model provides a unique view of the triaxial inertia tensor of nuclei.

Failures of the model reveal important low-energy hexadecapole degrees of freedom in nuclei.

Tutorial 2.2 Are nuclei axially symmetric? Case study: Os istopes. The video can be downloaded from https://doi.org/10.1088/978-0-7503-5643-5.

A View of Nuclear Data

Are nuclei axially asymmetric?
21. Unusual E2 properties

Some mixing effects of the model are "peculiar", and examples are realized in nuclei.

The model extends the "view" of E2 collectivity in nuclei beyond ground-state K = 0 bands.

Tutorial 2.3 Are nuclei axially symmetric? Unusual E2 properties. The video can be downloaded from https://doi.org/10.1088/978-0-7503-5643-5.

References

[1] Allmond J M, Zaballa R, Oros-Peusquens A M, Kulp W D and Wood J L 2008 Triaxial rotor model description of $E2$ properties in $^{186,\,188,\,190,\,192}$Os *Phys. Rev.* C **78** 014302

[2] Wood J L, Oros-Peusquens A M, Zaballa R, Allmond J M and Kulp W D 2004 Triaxial rotor model for nuclei with independent inertia and electric quadrupole tensors *Phys. Rev. C* **70** 024308

[3] Allmond J M, Wood J L and Kulp W D 2009 Triaxial rotor model description of quadrupole interference in collective nuclei: the P_3 term *Phys. Rev. C* **80** 021303

[4] Allmond J M, Wood J L and Kulp W D 2010 Destructive interference of $E2$ matrix elements in a triaxial rotor model *Phys. Rev. C* **81** 051305

[5] Allmond J M and Wood J L 2017 Empirical moments of inertia of axially asymmetric nuclei *Phys. Lett. B* **767** 226–31

[6] Davydov A S and Filippov G F 1958 Rotational states in even atomic nuclei *Nucl. Phys.* **8** 237–49

[7] Ayangeakaa A D *et al* 2023 Triaxiality and the nature of low-energy excitations in ^{76}Ge *Phys. Rev. C* **107** 044314

[8] Doherty D T *et al* 2017 Triaxiality near the ^{110}Ru ground state from Coulomb excitation *Phys. Lett. B* **766** 334–8

[9] Allmond J M 2013 Simple correlations between electric quadrupole moments of atomic nuclei *Phys. Rev. C* **88** 041307(R)

[10] Heyde K and Wood J L 2020 *Quantum Mechanics for Nuclear Structure: A Primer* (Bristol: IOP Publishing) https://iopscience.iop.org/book/mono/978-0-7503-2179-2

[11] Wu C Y *et al* 1996 Quadrupole collectivity and shapes of Os-Pt nuclei *Nucl. Phys. A* **607** 178–234

[12] Lim C S, Spear R H, Fewell M P and Gyapong G J 1992 Measurements of static electric quadrupole moments of the 2_1^+, 2_2^+, 4_1^+ and 6_1^+ states of ^{196}Pt *Nucl. Phys. A* **548** 308

[13] Heyde K and Wood J L 2020 *Quantum Mechanics for Nuclear Structure: An Intermediate Level View* (Bristol: IOP Publishing) https://iopscience.iop.org/book/mono/978-0-7503-2171-6

[14] Kulp W D, Allmond J M, Hatcher P, Wood J L, Loats J, Schmelzenbach P, Stapels C J, Krane K S, Larimer R-M and Norman E B 2006 Precision test of the rotor model from band mixing in ^{166}Er *Phys. Rev. C* **73** 014308

[15] Alaga G, Alder K, Bohr A and Mottelson B R 1955 Intensity rules for beta and gamma transitions to nuclear rotational states *Dan. Mat. Fys. Medd.* **29** 9

IOP Publishing

Nuclear Data
A collective motion view
David Jenkins and John L Wood

Chapter 3

How prevalent is shape coexistence in nuclei? Historical and closed-shell region views

Shape coexistence in nuclei is introduced from an historical perspective. The role of spectroscopic fingerprints—electromagnetic transition strengths (both enhanced and retarded, and electric monopole), nucleon transfer reaction spectroscopy, alpha-decay widths, nuclear masses (deduced two-neutron separation energies) and isotope shifts— is given in detail. The concept of intruder states, and their parabolic energy patterns and rotational bands is illustrated. Examples of mixing of coexisting structures are presented with quantitative detailed examples. Superdeformed bands are briefly illustrated.

Concepts: coexisting bands, particle–hole excitations, intruder states, $E0$ transitions and mixing.

Learning outcomes: the key data view from this chapter is the illustration of widespread occurrence of structures corresponding to more than one deformation ('shape') in a single nucleus. This is unequivocal in regions of closed shells wherein spherical structures are easily identified; thus, the presence of deformed structures stands out. The historical emergence of this feature has been arguably the most important development in our view of nuclear structure in the past 40 years, and this is illustrated in the selection of the data presented.

Shape coexistence in nuclei involves structures within a single nuclear species (A, Z, N) where at least two different deformations are manifested. Thus, we refer to the coexistence of spherical and deformed shapes or two different deformed shapes (even multiple shapes). It can also be referred to as shape 'isomerism': this term acknowledges the fact that the constituents (protons and neutrons) are the same but are 'arranged' in different ways.

The idea of shape coexistence originates in a paper by Morinaga in 1956 [1], i.e. in the earliest years of nuclear model building. However, it took decades to recognize that it is a widespread feature of nuclear structure. This evolution has been documented in three reviews [2–4] that directly evaluate the evidence for using the term 'shape coexistence'. There is also a focus issue of a journal that provides an on-going look from the perspective of various persons active in the exploration of nuclear shape coexistence [5], a more recent survey of experimental indications in new data [6], and a more recent look at electric monopole, $E0$ transition strength in nuclei and its association with shape coexistence [7]. The phenomenon can be described as having evolved from an exotic rarity, through behaviour confined to localized regions on the nuclear mass surface, to the likelihood that it occurs in all nuclei[1] (except the extreme lowest-mass nuclei).

The highest degrees of collectivity and deformation that have been observed in nuclei fall in the category of shape coexistence and have been termed 'super-deformed'. Note that this high degree of collectivity is not observed in association with nuclear ground states. The best example is illustrated in figure 1.16. A useful figurative guide is in terms of a solid spheroidal body (cf figure 1.4) and its major-to-minor axis ratio: it is 3:2 for the most-deformed ground-state structures of the rare-earth and actinide nuclei and it is 2:1 for the most superdeformed known structures in nuclei. It is unknown how many different deformations can be supported by a particular nuclear species; it is unknown what are the limitations to extreme nuclear deformation; it is unknown what variety of shapes are possible in nuclei, e.g. multipole moments or exotica such as 'banana-shapes'. In this chapter, an historical perspective is given to shape coexistence, its systematic features are presented, criteria used for inferring its presence are defined, and an outlook for exploration is given.

The most direct spectroscopic evidence that we possess for inferring the presence of deformation in nuclei is spectroscopic quadrupole moments and electric quadru-pole transition strengths, $B(E2)$. Such data can be reduced to an intrinsic quadru-pole moment, Q_0 for a rotational band. For non-zero deformation in nuclei, rotational bands can usually be identified. These are characterized by energy patterns which often serve as a proxy for the presence of deformation in a nucleus, i.e. when no electromagnetic data are available. Sometimes bands are recognizable by cascades of intra-band gamma rays. Often, when the band heads are above a few hundred keV, with the dominance of decays to low-lying states due to the E_γ^5 factor in observed $E2$ transition rates, intra-band cascades have not been characterized. Bands may be suggested via states that can only have one sensible band 'parentage'. Sometimes this parentage requires invoking an 'unexpected' structure, which is called an 'intruder' structure. The literature abounds with interpretations involving shape coexistence for which there is no direct experimental evidence of deformation, only such proxies. This becomes a serious problem when deformations are not large. Compounding this problem is that mixing between coexisting bands can occur. This can lead to complex inter-band decay patterns, even when the underlying structures are simple. Herein, we present the best evidence that we possess for nuclear shape

---

[1] A simple argument based on isobaric analog states leads to this conclusion: all isobaric sequences of nuclei contain multiple closed-shell nuclei and these will have spherical ground states; therefore, analogs of these states in open-shell nuclei, which will have deformed ground states, will constitute spherical coexisting structures. The idea is developed in [8].

coexistence. Patterns of occurrence are now becoming evident, and we introduce these patterns as a guide to future work.

3.1 Historical view of shape coexistence in nuclei

It is important to have an historical perspective on the emergence of the idea that nuclei can exhibit more than one shape in a single nuclear species. Particularly, it should be noted how many decades this required and how slow it was to arrive at a global perspective. It appears that this saga has been unique to the study of nuclear structure. Here, it is simply noted that one of the authors (JLW) has been engaged in this for 50 years and can share the view that the saga is likely only 'half told'.

Figure 3.1 shows the evidence for shape coexistence in the doubly closed shell nuclei ^{16}O and ^{40}Ca. It was ^{16}O that provided the inception of the idea of shape coexistence [1]; this was based on the indirect feature of excitations in ^{16}O that the lowest excited states should have negative parity because they would be particle–hole excitations. This expectation is based on a simple oscillator view of the nucleus wherein there is an alternating parity between shells. Thus, hole states in ^{16}O have negative parity and particle states have positive parity. The hole states are $1p_{1/2}$ and $1p_{3/2}$; the particle states are $2s_{1/2}$, $1d_{3/2}$ and $1d_{5/2}$. Recall that parity for shell model configurations is dictated as $(-1)^l$ and the parity of a 1p–1h excitation is the product of the parity of the particle and the hole. These negative-parity states are observed and can be matched to particle–hole excitations. But the lowest excited states in ^{16}O and ^{40}Ca have positive parity. The inference made by Morinaga [1] was that such states must involve excitations with at least $2\hbar\omega$ in energy, i.e. energy equal to two major shell gaps, either two nucleons across one shell gap or one nucleon across two shell gaps. To overcome the contradiction that multi-shell excitations are observed

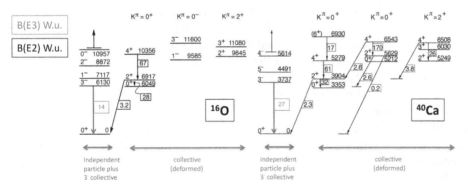

Figure 3.1. Excited states in the two $N = Z$, doubly closed shell nuclei, ^{16}O and ^{40}Ca. The decays of the 2^+_1 and 3^-_1 states to the ground state are characterized by weak collectivity as reflected in their $B(E2)$ and $B(E3)$ values. At the level of the first excited state, a 0^+ state in both nuclei, bands with highly collective character appear. These are shown with K quantum number assignments and $B(E2)$ values in Weisskopf units, where known. The states are partitioned into independent particle states, the 3^- collective states and the deformed collective states. Excitation energies are given in keV. See text for a scaled view of the $B(E2)$ values in these nuclei. The figure is reproduced from [9], copyright 2010 World Scientific Publishing Company.

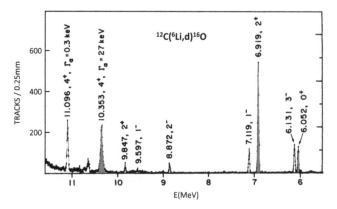

Figure 3.2. Spectrum of deuterons following the (^{6}Li,d) reaction on a ^{12}C target. The states which are interpreted as forming a deformed band in ^{16}O are highlighted in red. Other states populated are identified in figure 3.1. Energies of deuteron peaks are labelled in MeV for the corresponding states in ^{16}O. Reprinted figure with permission from [10]. Copyright 1974 by the American Physical Society.

lower in energy than single-shell excitations, Morinaga [1] suggested deformation, i.e. correlations are involved.

The spectroscopic evidence for deformed structures in ^{16}O and ^{40}Ca lies in the enormous[2] $B(E2)$ values between some of the states, as shown in figure 3.1. The spectroscopic evidence for multi-particle excitations being involved in these structures is provided by multi-nucleon transfer reactions, as shown in figure 3.2 [10], figure 3.3 [11] and figure 3.4 [12]. The key to making multi-particle-multi-hole interpretations of these states is the number of holes in each target nucleus with respect to the doubly closed shell and the number of nucleons added. The enormous energy 'cost' involved in exciting these configurations is countered by nearly equal enormous correlation energy gains in the binding energy. These correlations result in the deformation of the nucleus for these states. Thus, the states appear at low energy, as first excited states, with associated deformed bands.

The key paper in the history of the structures exhibited by ^{16}O and ^{40}Ca was by Brown and Green [13]. These two nuclei are iconic examples of shape coexistence in nuclei. Indeed, as shown in figure 3.1, ^{40}Ca possesses three coexisting structures: a spherical ground state, a deformed band built on the first excited state, and a superdeformed band built on the 0^+ state at 5.2 MeV. Note that these structures are $4\hbar\omega$ and $8\hbar\omega$ excitations, observed as essentially degenerate structures, at similar excitation energies to $1\hbar\omega$ excitations. The correlation energy gains involved are 50–100 MeV.

The language of 'particle' and 'hole' excitations, with respect to energy-shell structure based on the nuclear shell model, is prevalent in the discussion of shape coexistence in nuclei. Recall that the energy scale for shell structure in nuclei is

---

[2] Note that $B(E2)$ values scale as $A^{4/3}$: thus, 67 W.u. in ^{16}O scales to 1600 W.u. in ^{172}Yb (where the corresponding transition has a strength of 300 W.u.).

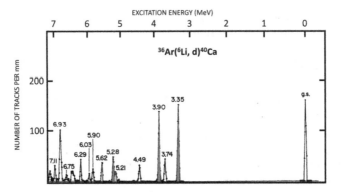

Figure 3.3. Spectrum of deuterons following the (^{6}Li,d) reaction on a ^{36}Ar target. The states which are interpreted as constituting a low-energy deformed band in ^{40}Ca are highlighted in red. The peaks highlighted in green are identified in figures 3.1 and 3.4. A summary view is provided in figure 3.1. Energies of deuteron peaks are labelled in MeV for the corresponding states in ^{40}Ca. Note: the peak labelled 5.28 MeV is a multiplet with components 5249 (2^+) and 5279 keV (4^+). Reprinted from [11] with permission from Elsevier.

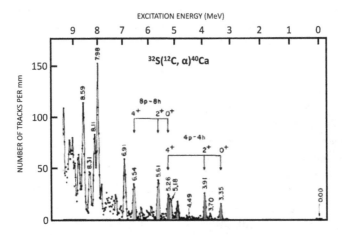

Figure 3.4. Spectrum of alphas following the (^{12}C, α) reaction on a ^{12}C target. The states which are interpreted as constituting deformed bands in ^{40}Ca are highlighted in red and green. Energies of alpha peaks are labelled in MeV for the corresponding states in ^{40}Ca. Members of two deformed bands, E_x (J $^\pi$): 3.353 (0^+), 3.904 (2^+), 5.279 (4^+), 6.930 (6^+), and E_x (J $^\pi$): 5.212 (0^+), 5.629 (2^+), 6.543 (4^+), 7.974 (6^+), energies in MeV, are assigned to these bands, cf figures 3.1 and 3.3. Note: the peak labelled 6.54 MeV is a multiplet with components 6.508 (4^+), 6.543 (4^+) and 6.582 (3^-); see also the caption to figure 3.3. Reprinted from [12] with permission from Elsevier.

quantified as $1\hbar\omega = 41A^{-1/3}$ MeV. The shell model provides an energy ordering that is basic to identifying many-particle configurations in spherical nuclei. It has a close analogy with the Periodic Table of the chemical elements and has given us the term 'aufbau' (German: to build up), as in electronic configurations of atoms of different elemental species. However, the aufbau process for ascertaining nucleon config-urations in nuclei does not follow the rules in atoms (e.g. Hund's rule). The reason is that the force between nucleons is attractive and further it is dominated by correlations. The result is that many shells can play a role in the low-energy configurations encountered in nuclei; further, these structures are generally deformed. At and close to closed shells, transfer reactions populate states with different deformations with dramatically different probabilities. It is important to recognize that transfer reactions probe the spherical components of generally a deformed structure, i.e. a structure which is a superposition of many spherical components. This is familiar in one-nucleon transfer reaction population of Nilsson bands in deformed odd-mass nuclei; it is less familiar, even unrecognized, in multi-nucleon transfer reaction population of coexisting bands at and near closed shells. This conceptual view will be revisited in this chapter.

It is perhaps amazing that the inception of the idea of shape coexistence did not kindle experimental programs to search for other examples of shape coexistence in nuclei. Rather, researchers 'stumbled' upon other examples; and there were no cross-references to other encounters of such interpretations. The result was that in isolated mass regions one found suggestions of shape coexistence. Two notable examples, in chronological order, were in ^{115,117}In [14] and in ^{199}Tl [15] (and see [16]). One shocking result changed this. An enormous isotope shift was observed between ^{185}Hg and ^{187}Hg [17], implying a sudden large change in the nuclear charge volume (likely due to a sudden change in ground-state deformation): it was sensational[3]. We present below first steps into a modern view of these historical landmark spectroscopic studies.

The current view of shape coexistence in ^{115,117,119}In is shown in figures 3.5, 3.6, and 3.7. In figure 3.5, the electromagnetic strengths dominate the evidence for enhanced (collective) transitions and retarded transitions (due to the large structural changes between the coexisting structures). Transfer reaction data, cf figure 3.6, reveal the dramatic difference between the coexisting structures, notably that the deformed structures have particle character coexisting with structures having hole character. A systematic view has been elucidated for candidate deformed bands in the odd-mass In isotopes and is shown for ^{115,117,119}In in figure 3.7.

Figures 3.5, 3.6 and 3.7 embody a rich variety of spectroscopic data which reach well beyond the context of the present focus on the historical first realization of shape coexistence in heavy nuclei. The most notable feature is the distinct character of three degrees of freedom that are important at low energy in the odd-mass In isotopes:

1. There are shell model proton hole states which are expected to dominate the In isotopes because the atomic number, $Z = 49$ locates them immediately 'below' a shell closure. Figure 3.6(b) shows the direct evidence for proton hole states as revealed by the ^{118}Sn(d,^{3}He)^{117}In one-proton pickup reaction.

---

[3] At least, it was for one of us (JLW) who was a first-year postdoctoral fellow and was looking for ideas to pursue. Yet further, many years later, in a conversation with the late Ernst Otten, the principal investigator on the Hg isotope shift work, JLW was told: 'the experiment was our last chance; we were informed that if we did not get a useful result our funding would be terminated.'

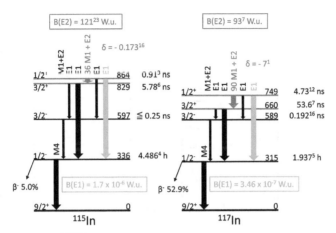

Figure 3.5. Intruder states in ^{115,117}In (levels highlighted in red) and the electromagnetic transition strengths that establish the shape coexistence associated with them. The 'intruder' nature reflects the impossibility of such states arising from a simple 2^+ core excited state coupling to the $9/2^+$ ground state, i.e. the $1/2^+$, $3/2^+$ spins must arise from a structure 'from somewhere else'. The figure resembles that in the original work [14] with only minor changes resulting from the most up-to-date information. The widths of the arrows qualitatively reflect the intensities of the transitions. Further details are discussed in the text. The juxtaposition of highly collective and highly retarded electromagnetic transitions, involving a common state, is a strong spectroscopic fingerprint of shape coexistence. The data are taken from ENSDF.

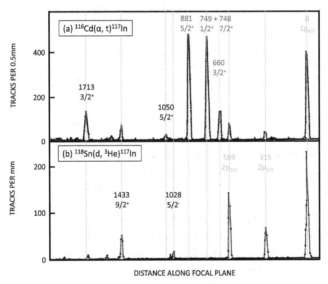

Figure 3.6. Spectroscopic proof of the 'particle' and 'hole' nature of low-lying states in ^{117}In. (a) Spectra for magnetically analyzed tritons from the ^{116}Cd(α,t)^{117}In reaction. The figure is adapted from one in [18], copyright (1972) by the American Physical Society. (b) Spectra for magnetically analyzed ^{3}He from the ^{118}Sn(d,^{3}He)^{117}In reaction. The figure is adapted from one in [19], copyright (1971), with permission from Elsevier. The peaks, corresponding to the ground state and excited states in ^{117}In, are labelled by the energies (in keV) of the states and by shell model configurations for spherical states and by spin-parities for deformed intruder states (cf figure 3.7). Further details are discussed in the text.

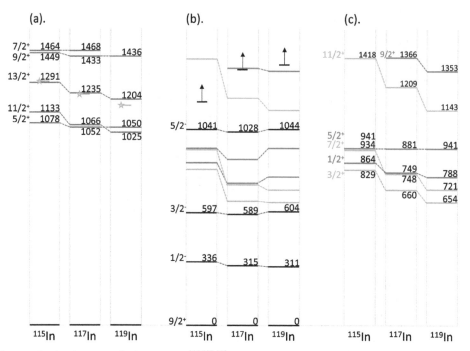

Figure 3.7. The lowest excited states in ^{115,117,119}In displayed by excitation mode (energies in keV): (a) Excitations assigned to the ^{116,118,120}Sn 'core' nuclei, i.e., the coupling $1g_{9/2}^{-1} \times 2_1^+$. The 2_1^+ energies are shown as orange stars. (b) Excitations assigned to the $2p_{1/2}^{-1}$ and $2p_{3/2}^{-1}$ shell model configurations, cf figure 3.6(b). (c) Excitations assigned to the intruder deformed bands, cf figure 3.6(a). Note: these energies are shown relative to the $5/2^+$ band members (their excitations relative to the ground states are shown in (b). Other states are shown in (b). Horizontal bars with vertical arrows indicate excitation energies above which states are omitted. Further details and discussion are given in the text. The data are taken from ENSDF.

2. The coupling of the proton hole to the first excited 2^+ state in the tin core with mass $A + 1$ is also anticipated and is depicted in figure 3.7(a).

3. The intruder state bands, which are an extended view of the present focus, can be interpreted as decoupled rotational bands built on the Nilsson configuration $1/2^+$ [431]. The decoupling separates the spin sequence $1/2^+$, $5/2^+$, $9/2^+$, $\cdots$ from $3/2^+$, $7/2^+$, $11/2^+$, $\cdots$ This is highlighted using different colours for these two spin sequences in figures 3.7(b) and (c). These details do not directly impact the interpretation of the intruder structures as deformed —that is manifested in the $B(E2)$ values shown in figure 3.5. Decoupled bands received attention in chapter 3 of [20], notably equation (3.16) and figure 3.12. Comparison of the spin sequence of the intruder band in ^{117}In, cf figure 3.7(c) with figure 3.12 in [20], notably the near degeneracy of the $1/2^+$ and $7/2^+$ states, shows that the decoupling parameter 'a' has the value -3 in ^{117}In. The 'fingerprint' manifested in the specific population of the intruder band states, as a function of spin in the ^{116}Cd$(\alpha, t)^{117}$In transfer reaction, is introduced in figure 3.23 in [20] and is briefly noted in section 5.1.2 of [20].

It is important to emphasize the distinct characters of the three noted structures in the odd-mass In isotopes. There is only isolated quantum mechanical mixing between the structures. The most pronounced feature is the fragmentation, due to mixing, of the $1g_{9/2}^{-1}$ hole strength between the ground state and the 1433 kcV excited state in ^{117}In evident in figure 3.6(b). The ^{116}Cd(α, t)^{117}In population of the proton hole states, $2p_{1/2}$ and $2p_{3/2}$, cf figure 3.6(a), indicates that the ^{116}Cd ground state (target nucleus) does not have a 'sharp' Fermi energy, as expected from pairing correlations, i.e. the ground-state of ^{116}Cd possesses some $2p_{1/2}$ and $2p_{3/2}$ 'hole' character. But the non-population of the intruder states by the ^{118}Sn(d, ^{3}He)^{117}In transfer reaction is remarkable and suggests that the ^{118}Sn ground state has a very sharp Fermi surface, as expected for a closed-shell nucleus.

The current view of shape coexistence in $^{193-201}$Tl is shown in figures 3.8 and 3.9. The Tl isotopes have received extensive study with respect to manifestation of shape coexistence. Indeed, multiple shape coexisting structures are emerging in the lighter Tl isotopes and this is discussed later. It is important to note that the In and Tl isotopes have a basic common feature: the lowest states in these isotopes are expected to be one-proton hole states with respect to the closed shells at $Z = 50$ and $Z = 82$, respectively.

There is a subtle and important difference between the intruder bands in the odd-mass In isotopes and the odd-mass Tl isotopes. The spin sequence in the In isotopes is 'irregular' due to the aforementioned decoupling, which is characteristic of a $K = 1/2$

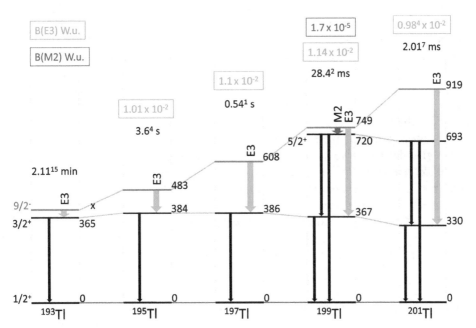

Figure 3.8. The $E3$ isomerism in $^{193-201}$Tl, which forms an important signature to the 9/2$^-$ intruder structure. The $B(E3)$ values in Weisskopf units are shown; these are additional support of the shape change involved. Energies are given in keV. The energy of the isomeric transition in ^{193}Tl is unknown but is estimated to be <13 keV. The figure is an updated view of a similar figure appearing in [15]. The data are taken from ENSDF.

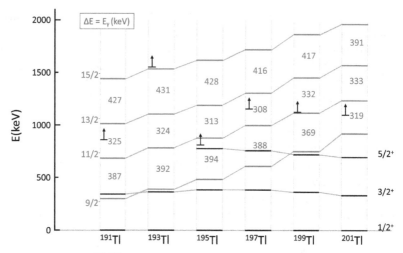

Figure 3.9. Band structure in $^{193-201}$Tl, which forms an important signature to the idea of shape coexistence. The figure is an updated view of a similar figure appearing in [16]. Level energy differences in keV are shown in blue. The data are taken from ENSDF.

band. The spin sequence in the Tl isotopes is 'regular', i.e., $9/2^-$, $11/2^-$, $13/2^-$, $\cdots$, i.e. it is characteristic of a $K = 9/2$ band. Intruder structures exhibit extreme values of Nilsson state Ω configurations, i.e. $\Omega = 1/2$ or $\Omega = j$ (where j characterizes the spherical shell model parent configuration underlying the Nilsson configuration). Thus, for prolate deformation an intruder state from the next higher shell is dominated by a Nilsson $\Omega = 1/2$ configuration and for oblate deformation an intruder state from the next higher shell is dominated by a Nilsson $\Omega = j$ configuration, cf figure 3.7 in [20]. Remarkably, the explanation of the intruder structures in the Tl isotopes is that they are a rare example of oblate deformation in nuclei.

In both the In and the Tl isotopes it is important to note the role of isomerism, as reflected in lifetimes of excited states, in establishing proof of the intruder character of the states involved. Further, the determination of transition multipolarities is critical in order to arrive at correct spins and parities of these states. This involves conversion electron spectroscopy, which is only performed in a few isolated programs of spectroscopic study of nuclei, worldwide. We particularly point to the conversion electron spectroscopy in these two advances in our understanding of nuclear structure: the In isotope studies were based on the legacy of the Siegbahn School in Uppsala[4]; the Tl isotopes studies were based on the pioneering work of Jack Hollander[5] at the Lawrence Berkeley Laboratory.

---

[4] The Siegbahn School refers to the work on x-ray spectroscopy by Manne Siegbahn in its application to atomic processes and then work by Kai Siegbahn in the application of electron spectroscopy to chemical analyses, ESCA. Both works were recognized with Nobel Prizes, in 1924 and 1981, respectively. These endeavours led to the development of electron spectroscopy for nuclear studies using magnetic spectrometers.
[5] Jack Hollander was the first person to apply semi-conductor spectrometers, in the form of lithium-drifted silicon detectors, to conversion-electron spectroscopy. This was then applied to both radioactive decay and in-beam spectroscopy studies of nuclear structure.

The original view of the large isotope shift in the Hg isotopes is shown in figure 3.10, the current view is shown in figure 3.11. The saga of shape coexistence in the Hg isotopes has been an example of one of the most thorough spectroscopic investigations across a broad mass range ever undertaken in nuclear structure

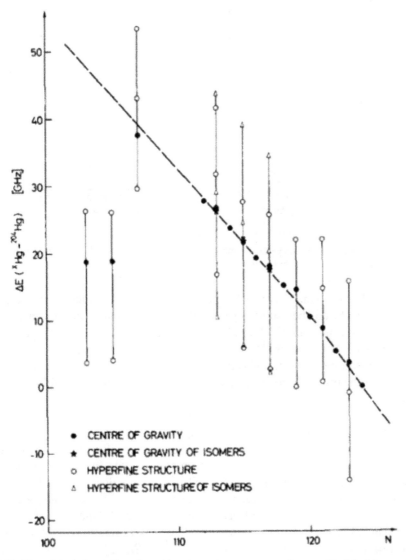

Figure 3.10. The original observation of a sudden change in mean-square charge radii of the Mercury isotopes, between ^{187}Hg and ^{185}Hg, which triggered the revolution in our view of shape coexistence in heavy nuclei. The data are for the hyperfine splitting of an optical transition, details of which can be found in the original study. Reprinted from [17], copyright (1972) with permission from Elsevier.

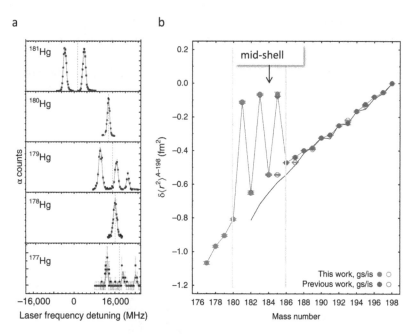

Figure 3.11. An up-to-date view of (a) the optical hyperfine structure observations in the very neutron-deficient mercury isotopes and (b) the isotopes shifts in the Mercury isotopes from $A = 177$ to $A = 198$. Reproduced from [21], copyright (2018) with permission from Springer Nature.

study[6]. It has been driven by compelling questions: 'how can nuclei just two protons removed from a closed shell suddenly exhibit a massive change in their mean-square charge radii (isotope shift)?' To take the perspective provided by figure 3.11, ^{181,182,185}Hg look 'as large as' ^{197}Hg. An additional puzzle was: 'why are only the odd-mass Hg isotopes showing this effect?' The even-mass isotopes looked 'normal'. At the time of the initial discovery, excited state data were not available for the light Hg isotopes.

The present view of excitations in the even-mass Hg isotopes is shown in figure 3.12. (Excitations in the relevant odd-mass Hg isotopes are presented in [4].) Two key historical follow-on studies of the even-mass Hg isotope shift were: an in-beam study of ^{184}Hg [22] which established via lifetime measurements that a rapid increase in deformation occurred at spin 4; and a ^{184}Tl radioactive decay study of ^{184}Hg [23] which established via conversion-electron measurements that a low-energy excited 0^+ state existed, and this was the head of a strongly deformed band.

Some years after the Mercury isotope shock, another one arrived. What appeared to be rotational bands were found in ^{112,114,116,118}Sn [24]. The current view of

---

[6] In defending a proposal, ca. 1975, to build on a successful study of excited states in ^{195}Hg by pursuing a systematic investigation of all the accessible odd-mass Hg isotopes, one of us (JLW) received the admonition that, since we understood one of these isotopes, there was no need to study the other ones. Unfortunately, the rebuttal: 'but we are studying the nuclear **many**-body problem', did not come to mind. This is likely one of the reasons that the saga has been such a long one.

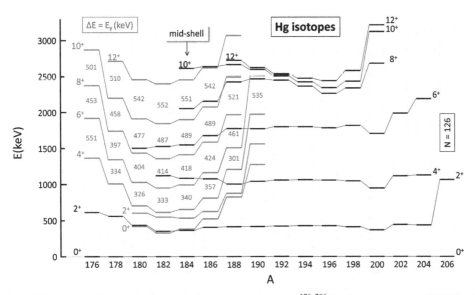

Figure 3.12. Systematics of excited states in the even-mass isotopes, [176-206]Hg. Intruder states are highlighted in red and energy differences between these states in keV are shown in blue. The data are taken from ENSDF.

[112,114,116,118]Sn is shown in figure 3.13. This includes the currently available $B(E2)$ data. Almost as great a shock came ten years after this with the discovery of the first superdeformed band in [152]Dy [25], which has already been a focal point herein (cf figure 1.16).

It then became evident that, with shape coexistence in indium and thallium isotopes and the above-noted common feature that they are adjacent to closed shells, a study of the lead isotopes was in order to see if the structure observed in the tin isotopes was matched (both being closed shells). This then followed [26, 27]. Indeed, it was recognized that an 'across-shell' connection could be established between coexisting structures in the thallium and the bismuth isotopes using alpha-decay spectroscopy [28]. The alpha-decay approach was extended yet further via the Po → Pb decays [29]. The current view of the even-mass Pb isotopes is shown in figure 3.14. The current views of the odd-mass alpha decay connections between the Bi and the Tl isotopes are shown in figure 3.15.

Yet another manifestation of shape coexistence emerged in the 1970s, with very subtle beginnings, but in the long run it has received considerable attention: the region centred on [32]Mg, i.e. at $N = 20$ and $Z = 12$. The first hints came from anomalies in masses of the sodium isotopes [30] and mean-square charge radii of sodium isotopes [31]; followed by observation of a low energy for the first-excited 2^+ state in [32]Mg [32]. The original sodium isotope results are shown in figure 3.16. The current view of excited state information in the $N = 20$ isotones is shown in figure 3.17.

To arrive at key features manifested in figure 3.17 took 40 years, with the discovery of excited 0^+ states in [32]Mg [33] and [34]Si [34]; and see [35], and a ground-state band

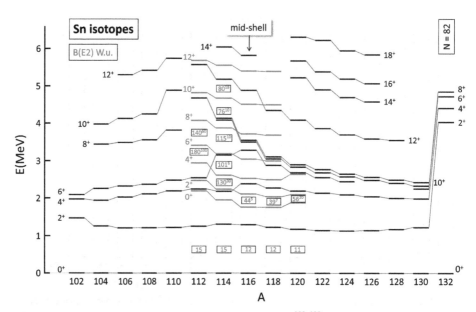

Figure 3.13. Systematics of excited states in the even-mass isotopes, $^{102-132}$Sn. Intruder states are highlighted in red and $B(E2)$ values in W.u. are shown in the black boxes in red. The $B(E2; 2_1^+ \to 0_1^+)$ values for $^{112-120}$Sn are shown. The data are taken from ENSDF.

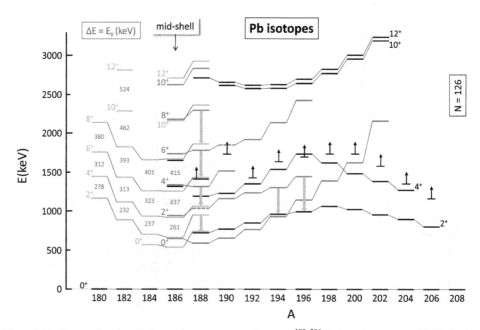

Figure 3.14. Systematics of excited states in the even-mass isotopes, $^{180-206}$Pb. Intruder states are highlighted in red and orange. Energy differences between intruder states in the most neutron-deficient isotopes are shown in blue. Selected $E0$ transitions are shown in orange. The data are taken from ENSDF. $B(E2)$ values are measured for the deformed band in ^{186}Pb and we direct the reader to consult ENSDF for their values.

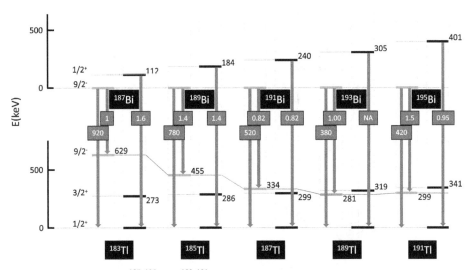

Figure 3.15. Alpha decays of $^{187-195}$Bi to $^{183-191}$Tl. The ground states of the Bi isotopes all correspond to the proton particle-configuration $1h_{9/2}$ and the ground states of the Tl isotopes all correspond to the proton-hole configuration $3s_{1/2}$. In each Bi isotope an alpha-decaying isomer with spin-parity $1/2^+$ is observed with fast, i.e. unhindered decay to the corresponding Tl ground state. Also, each Bi isotope ground state undergoes slow, i.e. hindered alpha decay to the corresponding Tl ground state and fast alpha decay to excited states with spin-parity $9/2^-$. The alpha decay hindrance factors are indicated (NA = not available). The $1/2^+$ isomers in the Bi isotopes are interpreted as intruder states related to the $3s_{1/2}$ configuration, but with 2p–1h character which possess deformation and corresponding collective bands, cf figure 4.16. The $9/2^-$ states in the Tl isotopes are interpreted as intruder states related to the $1h_{9/2}$, but with 1p–2h character which possess deformation and corresponding collective bands, cf figures 3.8, 3.9, 4.14, and 4.15. The data are taken from ENSDF. The figure is based on an original appearing in [28].

extending to spin 6 in ^{32}Mg [36]. This can be taken as a 'measure' of the difficulty of spectroscopic proof in neutron-rich nuclei 'far from stability'.

The near identical energy spacings of the ground-state band in ^{32}Mg and the highly deformed band in ^{38}Ar (cf figure 3.17), which is established as a neutron-pair excitation via two-neutron transfer reaction spectroscopy [37, 38] appears not to have been recognized. Evidently, there has been no interest in searching for a deformed band in ^{36}S. One finds frequent reference to the structure manifested in ^{32}Mg as a 'breakdown' or 'collapse' of the shell model: if one adopts this language, then the shell model breaks down at the excitation of the first excited state in ^{40}Ca; and there is the challenge of why shell model states are easily identified as coexisting structures, when the model is described as having collapsed.

There is the wider limitation of view resulting from language used in describing what is manifested in figure 3.17: namely, viewing the intruding structures centered on ^{32}Mg as an island ('island of inversion'); this may be why there has been no interest in ^{36}S. It appears that greater caution is needed in the language used in order not to mislead conceptualization of these structures or to limit their exploration. As such, the term 'island of inversion', while popular in the lexicon for discussing

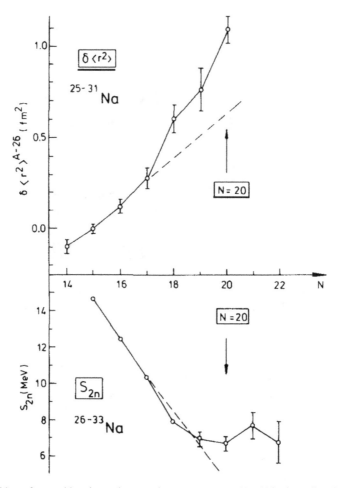

Figure 3.16. Evidence for a sudden change in ground-state structure at $N = 20$ in the sodium isotopes. Sodium two-neutron separation energies, S_{2n} are deduced from masses reported in [30]; sodium isotope shift data are deduced from laser induced hyperfine spectroscopy reported in [31].

intruder states and shape coexistence in nuclei, is a misnomer. Shape coexistence occurs widely and is not isolated in islands, as we present below.

With the above baseline we can turn to the current view of shape coexistence in nuclei. This is addressed in the next section, section 3.2 for nuclei at and adjacent to closed shells, and in section 4.1 for nuclei located in open-shell regions. In the following we provide a description of the most promising experimental approaches for a future exploration of shape coexistence in all nuclei. This includes some features that are not anticipated by the foregoing details. We complete the narrative with a look at some global perspectives in section 4.2 and some details of super-deformed bands in section 3.3.

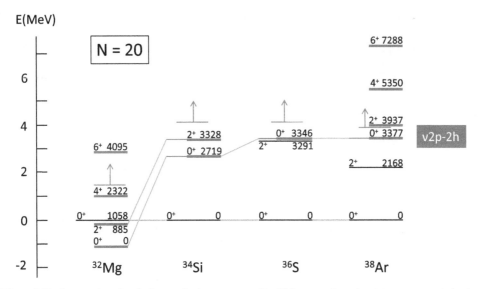

Figure 3.17. Systematics of excited states in the even-mass $N = 20$ isotones. Intruder states are suggested using red colouration for the relevant states. Energies are given in keV. The excited 0^+ state in ^{38}Ar at 3377 keV is a neutron-pair, $\nu 2p$–$2h$ excitation (see text), and the associated band is near-identical to the energy-spacing of the ground-state band in ^{32}Mg. Horizontal bars with vertically upwards-pointing arrows indicate excitations above which states are omitted. The data are taken from ENSDF and see text.

3.2 Signatures of shape coexistence for even–even nuclei in closed shell regions

The leading statement regarding a current view of coexisting shapes in nuclei at and near closed shells is that a large amount of work is needed to advance our understanding. But there are some powerful spectroscopic tools which can rapidly improve the status of nuclear shape coexistence. We illustrate these tools in order of likely useful results. We point to a very recent review that focusses on a roadmap for some immediate experimental exploration [6]. We strongly emphasize two aspects to the history of exploration of coexisting shapes in nuclei: first clues in a mass region are usually via indirect evidence (proxies); spectroscopic proof can take decades.

Electromagnetic transition strengths are the pre-eminent spectroscopic fingerprint of nuclear deformation, namely $B(E2)$ values and $E2$ matrix elements. There is a current revolution in multi-step Coulomb excitation with radioactive beams. A leading illustration is the Coulomb excitation of ^{182,184,186,188}Hg [39]. The gamma-ray spectrum from Coulomb excitation of ^{182}Hg is presented in figure 3.18. From the gamma-ray yields in Coulomb excitation, $E2$ matrix elements can be deduced, and these are given in table 3.1 for ^{182}Hg. Further, these data point beyond $E2$ matrix elements to the role of electric monopole, $E0$ transition strengths in association with nuclear shape coexistence. These are shown for ^{182,184,186}Hg in figure 3.19, together with $B(E2)$ values in Weisskopf units. Note that the $E0$ transitions were established earlier by conversion-electron spectroscopy; but deduction of $E0$ strengths requires

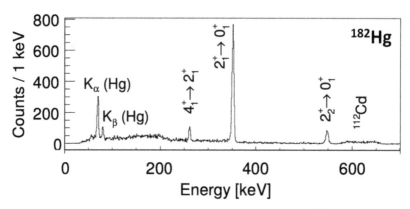

Figure 3.18. Gamma-ray spectrum from multi-step Coulomb excitation of ^{182}Hg. This required the re-acceleration of a radioactive ion beam, carried out using the REX-ISOLDE facility. The feature labelled as '^{112}Cd' is a Doppler broadened peak due to Coulomb excitation of the target by the radioactive Hg beam. Reprinted from [39] under CC-BY-4.0 license.

Table 3.1. $E2$ matrix elements in eb for ^{182}Hg determined by multi-step Coulomb excitation, cf figure 3.18. The entry in square brackets means $-2.2 \leqslant \langle \; |E2| \; \rangle < 0.9$. The data are taken from [39]. These matrix elements can be converted into $B(E2)$ values using equation (1.4) and $1 \; e^2b^2 = 163$ W.u., cf figure 3.19.

$\langle I_i||E2||I_f \rangle$	Matrix element (eb)
$\langle 0_1^+||E2||2_1^+ \rangle$	1.29(4)
$\langle 2_1^+||E2||4_1^+ \rangle$	3.70(6)
$\langle 0_1^+||E2||2_2^+ \rangle$	−0.6(1)
$\langle 0_2^+||E2||2_1^+ \rangle$	[−2.2, 0.9]
$\langle 0_2^+||E2||2_2^+ \rangle$	−1.25(30)
$\langle 2_1^+||E2||2_2^+ \rangle$	−2.0(3)
$\langle 2_2^+||E2||4_1^+ \rangle$	3.3(4)

lifetime data, which only became available when the $E2$ matrix elements were determined (see [7]).

A similar detailed view of shape coexistence to that in the Hg isotopes has recently been obtained for ^{42}Ca [40]. Details are shown in figure 3.20. The spectroscopic data available for ^{42}Ca permit detailed calculations to be made, these are shown in figure 3.21. Of particular note is the emergence of the large transition strength between the deformed and spherical states manifested in $B(E2; 0_2 \rightarrow 2_1) = 57$ W.u. This arises entirely from mixing, i.e. there is *zero intrinsic strength* involved.

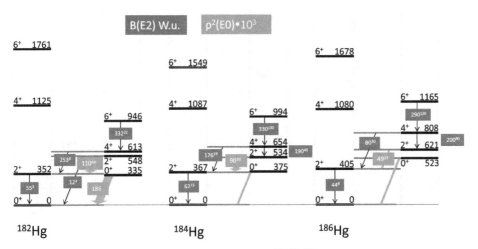

Figure 3.19. Selected $E2$ and $E0$ transition strengths in ^{182,184,186}Hg, given as $B(E2)$s in W.u and $\rho^2(E0) \times 10^3$ values. The data are taken from [39], ENSDF, and table 3.1. Reprinted from [7], copyright (2022) with permission from Elsevier.

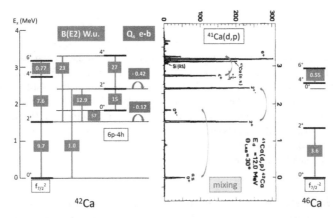

Figure 3.20. The nucleus ^{42}Ca viewed from the perspectives of multi-step Coulomb excitation and one-neutron transfer reactions. On the left are $B(E2)$ values in Weisskopf units and diagonal $E2$ matrix elements in eb deduced from the Coulomb excitation measurements [40]. In the centre is the proton spectrum resulting from the ^{41}Ca(d,p)^{42}Ca one-neutron transfer reaction [41]. The fragmentation of transfer strength is indicated by the arrows with respect to the population of two states each of spin-parity 0^+, 2^+, and 4^+, where only one state of each spin-parity would be populated for a seniority $\nu 1f_{7/2}^2$ dominated structure in ^{42}Ca. On the right is the (near) pure $\nu 1f_{7/2}^{-2}$ dominated structure in ^{46}Ca. The fragmentation of transfer strength is explained by configuration mixing between the seniority structure and the deformed structure, details of which are presented in figure 3.21. Other data are taken from ENSDF. Reprinted from [42] under CC-BY-4.0 license.

A subtle issue is one of where to focus study using the powerful technique of multi-step Coulomb excitation. We raise this issue because the overwhelming drive within the nuclear physics community is towards rare isotope studies, with respect to which ^{182,184,186}Hg are examples, and ^{42}Ca is not. Nevertheless, both types of study

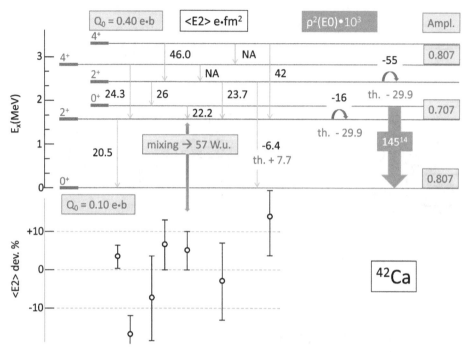

Figure 3.21. A two-state mixing approach to the description of the low-energy states in ^{42}Ca. The mixing calculations are carried out, independently, for spins 0, 2, and 4. The mixing amplitudes are fixed empirically from the fragmentation of the one-neutron transfer strengths and are given on the right-hand side of the figure. There are two fitted parameters for the $E2$ properties: $Q_0 = 0.10$ eb for the spherical configurations and $Q_0 = 0.40$ eb for the deformed configurations. For the deformed configurations, the Q_0 value is multiplied by the rotor model Clebsch–Gordan coefficients, viz. $-1.195Q_0$ for $2_1 - 2_1$ and $1.604Q_0$ for $2_1 - 4_1$. The differences between theory and experiment are shown as $\{(\langle E2 \rangle_{ex} - \langle E2 \rangle_{th})/\langle E2 \rangle_{ex}\} \times 100\%$. For other details, see text. Reprinted from [42] under CC-BY-4.0 license.

are essential. We state the adage: 'do not build castles on sand'. Rare isotope studies are limited to very low event rates, and this leads to substantial interpretations with a lack of spectroscopic detail; even model-inspired assignments of spins and parities followed by interpretations using the same model, which guarantee that the model cannot be falsified. What is critical is to be aware of the global features of nuclear structure, and we will emphasize this herein.

Electric monopole transition strengths have been recognized to be a useful indicator of the presence of shape coexistence when there is mixing of configurations with different mean-square charge radii [43]. This is manifested in figure 3.19. Other examples are ^{110,112,114}Cd, shown in figure 3.22, and ^{114,116,118,120}Sn (cf figure 3.13), shown in figure 3.23. The $E2$ transition strengths that support shape coexistence are included in figure 3.22 for the Cd isotopes and are shown in figure 3.24 for the Sn isotopes. Note again that the cadmium isotopes are two-proton holes away from the $Z = 50$ closed shell and the mercury isotopes are two-proton holes away from the

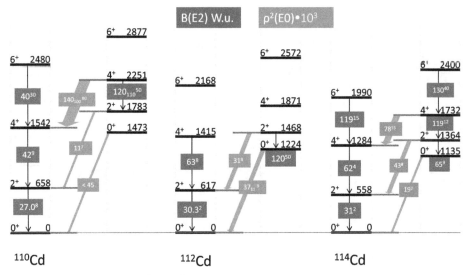

Figure 3.22. Shape coexistence in ^{110,112,114}Cd and associated $E2$ and $E0$ transition strengths given as $B(E2)$s in W.u. and $\rho^2(E0) \times 10^3$ values. The data for ^{110}Cd are taken from [46]. Other data are taken from ENSDF. Reprinted from [7], copyright (2022) with permission from Elsevier.

$Z = 82$ closed shell[7]. We point to a very recent review [7] on $E0$ transition strengths in nuclei, which follows on from an earlier work [45] wherein the association between $E0$ transition strengths, mean-square charge radii, and shape coexistence is explored. Key details of the mixing mechanism by which $E0$ strength is generated are given in chapter 4 of [20], and see below.

A distinct and complementary spectroscopic view of shape coexistence in nuclei, especially at and near closed shells, is provided by two-nucleon transfer reaction spectroscopy. Two examples, for ^{118}Sn and ^{110}Cd, are shown in figure 3.25. Such spectroscopy reveals the pair-excitation structure of the deformed states which is based on the hole character of the target nuclei used. This leads to a direct view of the 'enhanced' nucleon configuration space underlying the more-deformed structures. A simplified view of this is depicted in figure 3.26. The view immediately suggests that the cadmium structures should be mirrored by such structures in the tellurium isotopes. States in ^{118}Te are compared with its isotone ^{114}Cd in figure 3.27.

The manifestation of shape coexistence in nuclei at and near $N = 50$ and $N = 82$ has received essentially no attention. Two-neutron transfer-reaction spectroscopy clearly identifies candidate structures, as shown in figures 3.28 and 3.29. However, there are no electromagnetic data to indicate deformation in association with these structures. An explanation of this is the higher excitation energies of the states involved: to study states at these higher excitations involves complex spectroscopy due to the level density which will be high.

---

[7] Shape coexistence was first suggested in the cadmium isotopes in ^{110}Cd [44], but it was not recognized as a structure matched in the Hg isotopes.

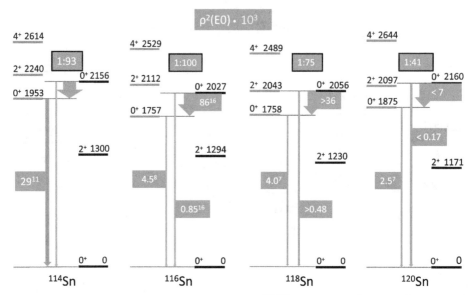

Figure 3.23. The manifestation of $E0$ transition strength in $^{114-120}$Sn, shown as $\rho^2(E0) \times 10^3$ values. Details are discussed in the text. Reprinted from [7], copyright (2022) with permission from Elsevier.

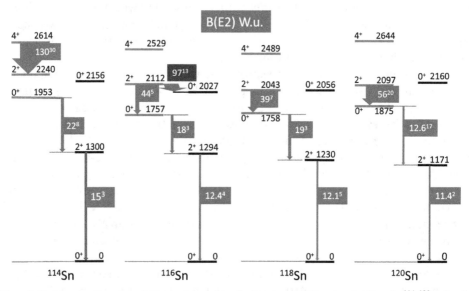

Figure 3.24. The manifestation of $E2$ transition strength, shown as $B(E2)$ values in W.u., in $^{114-120}$Sn for the same states as shown in figure 3.23. Reprinted from [7], copyright (2022) with permission from Elsevier.

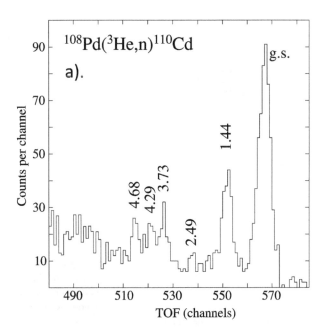

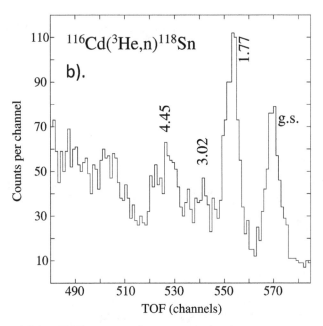

Figure 3.25. Time-of-flight (TOF) spectra for neutrons in the two-neutron transfer reactions: (a) ^{108}Pd(^{3}He,n)^{110}Cd and (b) ^{116}Cd(^{3}He,n)^{118}Sn. The events corresponding to the ground states are marked as g.s. All the strong peaks correspond to 0^+ states. The peaks at 1.44 MeV in ^{110}Cd and 1.77 MeV in ^{118}Sn match the excitation energies of the lowest energy deformed states in these isotopes. The spectra are taken from [47]. The figure was made available to us, courtesy of Paul Garrett.

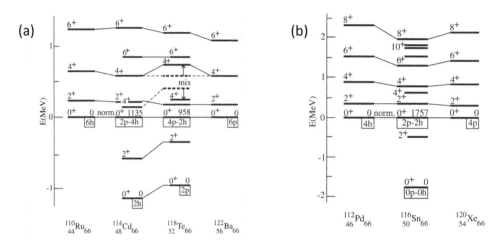

Figure 3.26. A view of shape coexistence near $Z \sim 50$ in terms of intruder-state configurations and their dominant particle–hole structure. In (a) the $\pi(2p–4h)$ and $\pi(4p–2h)$ structures in ^{114}Cd and ^{118}Te, respectively, are compared to the ground-state structures in ^{110}Ru $\pi(6h)$ and ^{122}Ba $\pi(6p)$. Note the possibility of mixing in ^{118}Te. In (b) the $\pi(2p–2h)$ structure in ^{116}Sn is compared to the ground-state structures in ^{112}Pd $\pi(4h)$ and ^{120}Xe $\pi(4p)$. All these nuclei have $N = 66$. The figures are similar to ones appearing in [4] and are based on figures appearing in [48].

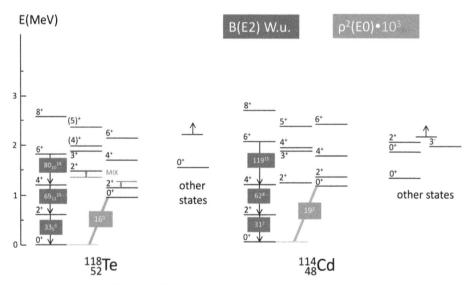

Figure 3.27. The isotones ^{114}Cd and ^{118}Te and a comparison of selected $E2$ strengths, $B(E2)$ in W.u. and $E0$ transition strengths given as $\rho^2(E0) \times 10^3$ values. These nuclei are depicted in figure 3.26(a) and can be viewed as 'particle' and 'hole' mirrors of each other. Evidently the location of these two nuclei in different shells has little impact on their basic collective character. This suggests that the effective interactions between nucleons, with respect to the basic collectivity of nuclei at low energy, is shell independent. Reprinted from [7], copyright (2022) with permission from Elsevier.

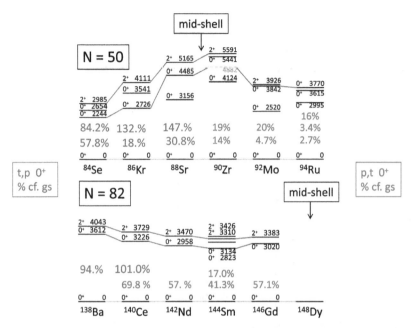

Figure 3.28. Evidence for $\nu(2p\text{–}2h)$ states in the $N = 50$ and $N = 82$ isotones. These structures are identified by two-neutron transfer reactions: the (t,p) reactions [49–51] and (p,t) reactions [52–54], at $N = 50$; the (t,p) reactions [55, 56], and (p,t) reactions [56–58] at $N = 82$. The population of excited 0^+ states is given relative to the ground states as a percentage. The (t,p) data are shown in blue and the (p,t) data are shown in red. Thus, the ^{82}Se(t,p)^{84}Se reaction populates the 0^+ state in ^{84}Se at 2244 keV with 57.8% the population of the 0^+ ground state. Dashed lines are to guide the eye and do not imply corresponding structures. Population data for the 2^+ states are not given but can be found in the data sources given.

The manifestation of shape coexistence at and near closed shells in light nuclei is addressed further in section 4.2 (figure 3.20).

3.3 Superdeformed bands

Superdeformed bands have been encountered in the foregoing narrative. There is a vast literature[8] describing such bands in nuclei, ranging across the entire mass surface. In the details, one finds that only in a handful of cases have these bands of states been 'connected' to low-lying states. The transitions involved are often termed 'draining' transitions. The lack of connecting transitions means that these bands 'float' with a lack of absolute excitation and absolute spin-parity information. Most assignments of spin-parities to superdeformed band heads have been made using model-based arguments.

The draining of superdeformed bands, where it has been elucidated, is extremely complex. There are many decay paths and so data of very high statistical quality are needed to see even the strongest decay paths. An early, successful 'mapping' of the

---

[8] There is a compendium of superdeformed band data available, published as a report [61] and as an issue of *Nuclear Data Sheets* [62].

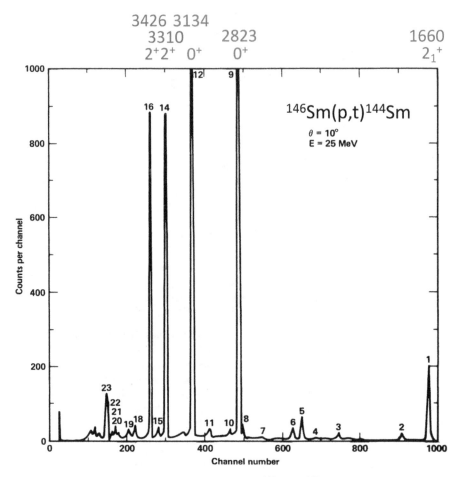

Figure 3.29. Spectrum of tritons following the reaction ^{146}Sm(p,t)^{144}Sm, cf figure 3.28. The figure is reproduced from [58], copyright (1983) by the American Physical Society. Evidently there is strong mixing between the intruder configurations and other configurations, manifested in the fragmentation of transfer strength to the 0^+ and 2^+ states.

strongest decay paths (all <1%) was in ^{152}Dy, and details are shown in figure 3.30. This nucleus has been used herein as a 'poster' example of a superdeformed band. If the band is interpreted as possessing $K = 0$, the associated 0^+ band head lies at an excitation of about 7 MeV.

Superdeformed bands are manifested as so-called fission isomers in the actinide region. An illustration of such structure is depicted in figure 3.31. The draining of such an isomer was elucidated in ^{238}U [60].

The lack of excitation energies, spin-parities, and K quantum numbers for superdeformed bands, independent of the model-dependent assignments, places this large body of nuclear data beyond discussion in the present approach.

A global view of the occurrence of these (high-energy, high-spin) superdeformed bands is presented in figure 3.32.

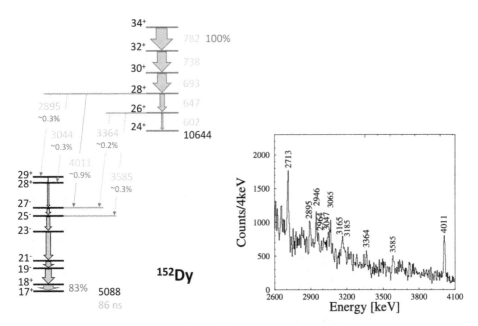

Figure 3.30. Draining of the superdeformed, SD band in ^{152}Dy. The draining transitions are double-coincidence gated on γ rays in the SD band and γ rays de-exciting the 86 ns isomer at 5088 keV, to which the observed feeding is 83% of the SD-band population. The conclusion is that the draining is spread over a very large number of paths that are too weak to be characterized at this level of statistical quality. More details of the SD band are given in 1.16. The level scheme presented is based on details appearing in [59]. Reprinted figure with permission from [59]. Copyright (2002) by the American Physical Society.

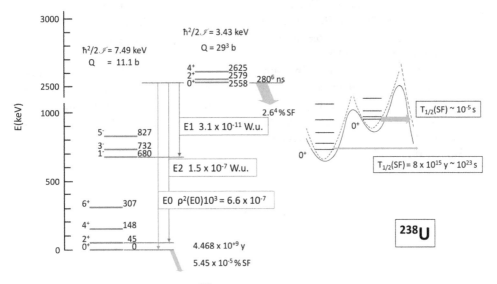

Figure 3.31. Details of the fission isomer in ^{238}U. Note the spontaneous fission partial half lives are in the ratio $1:10^{28}$; this is a consequence of the different barrier heights and widths and the effect on tunnelling for the separating fission fragments. The data are taken from ENSDF.

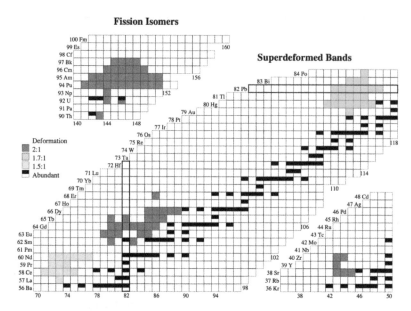

Figure 3.32. Global view of the occurrence of superdeformed structures in nuclei. The figure is reproduced from [61] and is similar to one appearing in [62].

3.4 Exercises

3-1. At the beginning of this chapter, ^{16}O and ^{40}Ca, were introduced as clear examples of shape coexistence manifested in doubly closed-shell nuclei e.g. see figure 3.1 *et seq.* Consider whether the doubly closed-shell nucleus ^{56}Ni ($N = Z = 28$) exhibits similar structure. Using figure 3.1 as a guide, organize the available data for ^{56}Ni into a similar pattern, taking the data from ENSDF.

3-2. The nuclei ^{113,119}In exhibit similar structure i.e. shape coexistence as noted earlier in the chapter in respect of ^{115}In and ^{117}In. Using figure 3.5 as a guide, organize the data for ^{113,119}In into a similar pattern, taking the data from ENSDF. In particular, identify enhanced $E2$ transitions and retarded $E1$ transitions in these nuclei.

3-3. Make similar figures to figure 3.26 for ^{112,114}Cd, ^{114,118}Sn.

A View of Nuclear Data

Nuclear shape coexistence
6. Some global features of shape coexistence

An up-to-date view of excitation patterns involving shape coexistence in nuclei is presented.

The dominant pattern is a "parabolic" energy trend with a minimum near mid-shell points.

Evidence for more than one coexisting shape appears in many nuclei.

Tutorial 3.1 Some global features of shape coexistence. The video can be downloaded from https://doi.org/10.1088/978-0-7503-5643-5.

A View of Nuclear Data

Nuclear shape coexistence
7. Ground-state intruders and coexistence

The characteristic "parabolic" energy patterns associated with coexisting nuclear shapes result in the "intrusion" of strongly deformed structures to become ground states in regions of weak or zero deformation.

These intrusions lead to dramatic changes in ground-state properties
--spin, magnetic moment, quadrupole moment, mean-square charge radius—
such properties are widely mapped using optical hyperfine structure in atoms containing these nuclei.

Tutorial 3.2 Ground state intruders and coexistence. The video can be downloaded from https://doi.org/10.1088/978-0-7503-5643-5.

A View of Nuclear Data

Experimental probes
11. Coulomb excitation, multi-step

A basic view of multi-step Coulomb excitation of excited states in nuclei is presented.

Patterns in strongly deformed nuclei (rotations) and weakly deformed nuclei are illustrated.

Tutorial 3.3 Multi-step Coulomb excitation. The video can be downloaded from https://doi.org/10.1088/978-0-7503-5643-5.

A View of Nuclear Data

Mixing of simple model degrees of freedom
15. Mixing of spherical and deformed structures: ^{42}Ca

A simple two-state mixing analysis is illustrated

The analysis achieves quantitative agreement with E2 data at the "10%" level

The mixing analysis reveals that more complex models are inferior

Tutorial 3.4 Mixing of spherical and deformed structures: ^{42}Ca. The video can be downloaded from https://doi.org/10.1088/978-0-7503-5643-5.

References

[1] Morinaga H 1956 Interpretation of some of the excited states of 4n self-conjugate nuclei *Phys. Rev.* **101** 254

[2] Heyde K, Van Isacker P, Waroquier M, Wood J L and Meyer R A 1983 Coexistence in odd-mass nuclei *Phys. Repts.* **102** 291

[3] Wood J L, Heyde K, Nazarewicz W, Huyse M and Van Duppen P 1992 Coexistence in even-mass nuclei *Phys. Repts.* **215** 101

[4] Heyde K and Wood J L 2011 Shape coexistence in atomic nuclei *Rev. Mod. Phys.* **83** 1467

[5] Wood J L and Heyde K 2016 A focus on shape coexistence in nuclei *J. Phys.* G **43** 020402

[6] Garrett P E, Zielińska M and Clément E 2022 An experimental view of shape coexistence in nuclei *Prog. Part. Nucl. Phys.* **124** 103931

[7] Kibédi T, Garnsworthy A B and Wood J L 2022 Electric monopole transitions in nuclei *Prog. Part. Nucl. Phys.* **123** 103930

[8] Rowe D J and Wood J L 2018 A relationship between isobaric analogue states and shape coexistence in nuclei *J. Phys.* G **45** 06LT01

[9] Rowe D J and Wood J L 2010 *Fundamentals of Nuclear Models: Foundational Models* (Singapore: World Scientific)

[10] Debevec P T, Fortune H T, Segel R E and Tonn J F 1974 Anomalous population of two 4^+ states in ^{12}C(^{6}Li,d)^{16}O *Phys. Rev.* C **9** 2451

[11] Betts R R, Fortune H T, Bishop J N, Al-Jadir M N I and Middleton R 1977 Alpha transfer to 4p-4h states in ^{40}Ca *Nucl. Phys.* A **292** 281–7

[12] Middleton R, Garrett J D and Fortune H T 1972 Search for multiparticle-multihole states of ^{40}Ca with the ^{32}S(^{12}C,α) reaction *Phys. Lett.* B **39** 339–42

[13] Brown G E and Green A M 1966 Even parity states of ^{16}O and ^{17}O *Nucl. Phys.* **75** 401–17

[14] Bäcklin A, Fogelberg B and Malmskog S G 1967 Possible deformed states in ^{115}In and ^{117}In *Nucl. Phys.* A **96** 539–60

[15] Newton J O, Cirilov S D, Stephens F S and Diamond R M 1970 Possible oblate shape of $9/2^-$ isomer in ^{199}Tl *Nucl. Phys.* A **148** 593–614

[16] Newton J O, Stephens F S and Diamond R M 1974 Rotational bands in the light odd-mass Tl nuclei *Nucl. Phys.* A **236** 225–51

[17] Bonn J, Huber G, Kluge H-J, Kugler L and Otten E W 1972 Sudden change in the nuclear charge distribution of very light mercury isotopes *Phys. Lett.* B **38** 308–11

[18] Harar S and Horoshko R N 1972 Study of the level scheme of ^{117}In via proton transfer reactions *Nucl. Phys.* A **183** 161–72

[19] Weiffenbach C V and Tickle R 1971 Structure of odd-A indium isotopes determined by the (d,^{3}He) reaction *Phys. Rev.* C **3** 1668

[20] Jenkins D G and Wood J L 2021 *Nuclear Data: A Primer* (Bristol: IOP Publishing) https://iopscience.iop.org/book/mono/978-0-7503-2674-2

[21] Marsh B A *et al* 2018 Characterization of the shape-staggering effect in mercury nuclei *Nat. Phys.* **14** 1163–7

[22] Rud N, Ward D, Andrews H R, Graham R L and Geiger J S 1973 Lifetimes in the ground-state band of ^{184}Hg *Phys. Rev. Lett.* **31** 1421

[23] Cole J D *et al* 1976 Behavior of the excited deformed band and search for shape isomerism in ^{184}Hg *Phys. Rev. Lett.* **37** 1185

[24] Bron J *et al* 1979 Collective bands in even mass Sn isotopes *Nucl. Phys.* A **318** 335–51

[25] Twin P J *et al* 1986 Observation of a discrete-line superdeformed band up to 60 $\hbar$ in ^{152}Dy *Phys. Rev. Lett.* **57** 811

[26] Van Duppen P, Coenen E, Deneffe K, Huyse M, Heyde K and Van Isacker P 1984 Observation of low-lying $J^\pi = 0^+$ states in the single-closed-shell nuclei $^{192-198}$Pb *Phys. Rev. Lett.* **52** 1974

[27] Van Duppen P, Coenen E, Deneffe K, Huyse M and Wood J L 1987 β^+/electron capture decay of ^{192,194,196,198,200}Bi: Experimental evidence for low lying 0^+ states *Phys. Rev. C* **35** 1861

[28] Coenen E, Deneffe K, Huyse M, Van Duppen P and Wood J L 1985 α Decay of neutron-deficient odd Bi nuclei: shell-model intruder states in Tl and Bi isotopes *Phys. Rev. Lett.* **54** 1783

[29] Van Duppen P, Coenen E, Deneffe K, Huyse M and Wood J L 1985 Low-lying $J^\pi = 0^+$ states in ^{190,192}Pb populated in the α-decay of ^{194,196}Po *Phys. Lett. B* **154** 354–7

[30] Thibault C, Klapisch R, Rigaud C, Poskanzer A M, Prieels R, Lessard L and Reisdorf W 1975 Direct measurement of the masses of ^{11}Li and $^{26-32}$Na with an on-line mass spectrometer *Phys. Rev. C* **12** 644

[31] Huber G *et al* 1978 Spins, magnetic moments, and isotope shifts of $^{21-31}$Na by high resolution laser spectroscopy of the atomic D_1 line *Phys. Rev. C* **18** 2342

[32] Détraz C, Guillemaud D, Huber G, Klapisch R, Langevin M, Naulin F, Thibault C, Carraz L C and Touchard F 1979 Beta decay of $^{27-32}$Na and their descendants *Phys. Rev. C* **19** 164

[33] Wimmer K *et al* 2010 Discovery of the shape coexisting 0^+ state in ^{32}Mg by a two neutron transfer reaction *Phys. Rev. Lett.* **105** 252501

[34] Rotaru F *et al* 2012 Unveiling the intruder deformed 0_2^+ state in ^{34}Si *Phys. Rev. Lett.* **109** 092503

[35] Lica R *et al* 2019 Normal and intruder configurations in ^{34}Si populated in the β decay of ^{34}Mg and ^{34}Al *Phys. Rev. C* **100** 034306

[36] Crawford H L *et al* 2016 Rotational band structure in ^{32}Mg *Phys. Rev. C* **93** 031303

[37] Flynn E R, Hansen O, Casten R F, Garrett J D and Ajzenberg-Selove F 1975 The (t,p) reaction on ^{36,38,40}Ar *Nucl. Phys. A* **246** 117–40

[38] Miura K, Hiratate Y, Shoji T, Suelhro T and Ohnuma H 1980 The ^{40}Ar(p,t)^{38}Ar reaction at 52 MeV *Nucl. Phys. A* **334** 389–400

[39] Wrzosek-Lipska K *et al* 2019 Electromagnetic properties of low-lying states in neutron-deficient Hg isotopes: Coulomb excitation of ^{182}Hg, ^{184}Hg, ^{186}Hg and ^{188}Hg *Eur. Phys. J. A* **55** 130

[40] Hadyńska-Klek K *et al* 2018 Quadrupole collectivity in ^{42}Ca from low-energy Coulomb excitation with AGATA *Phys. Rev. C* **97** 024326

[41] Ellegaard C, Lien J R, Nathan O, Sletten G, Ingrebretsen F, Osnes E, Tjøm P O, Hansen O and Stock R 1972 The $(1f_{7/2})^2$ multiplet in ^{42}Ca *Phys. Lett. B* **40** 641–4

[42] Stuchbery A E and Wood J L 2022 To shell model, or not to shell model, that is the question *Physics* **4** 697–773

[43] Kantele J *et al* 1979 Absolute $E0$ and $E2$ transition rates and collective states in ^{116}Sn *Z. Phys. A* **289** 157

[44] Meyer R A and Peker L 1977 Evidence for coexistence of shapes in even-mass Cd nuclei *Z. Phys. A* **283** 379–82

[45] Wood J L, Zganjar E F, De Coster C and Heyde K 1999 Electric monopole transitions from low energy excitations in nuclei *Nucl. Phys. A* **651** 323

[46] Jigmeddorj B *et al* 2016 Conversion electron study of ^{110}Cd: evidence of new $E0$ branches *Eur. Phys. J. A* **52** 36

[47] Fielding H W *et al* 1977 0$^+$ states observed in Cd and Sn nuclei with the (^{3}He,n) reaction *Nucl. Phys. A* **281** 389

[48] Heyde K, De Coster C, Jolie J and Wood J L 1992 Intruder analog states: new classification of particle-hole excitations near closed shells *Phys. Rev. C* **46** 541

[49] Mullins S M, Watson D L and Fortune H T 1988 ^{82}Se(t,p)^{84}Se reaction at 17 MeV *Phys. Rev. C* **37** 587

[50] Flynn E R, Sherman J D, Stein N, Olsen D K and Riley P J 1976 ^{84}Kr(t,p)^{86}Kr and ^{86}Kr(t,p)^{88}Kr reactions *Phys. Rev. C* **13** 568

[51] Ragaini R C, Knight J D and Leland W T 1970 Levels of ^{88}Sr from the ^{86}Sr(t,p)^{88}Sr reaction *Phys. Rev. C* **2** 1020

[52] Ball J B, Auble R L and Roos P G 1971 Study of the Zirconium isotopes with the (p,t) reaction *Phys. Rev. C* **4** 196

[53] Ball J B and Larsen J S 1972 Systematics of $L = 0$ transitions observed in the (p,t) reaction of nuclei near $N = 50$ *Phys. Rev. Lett.* **29** 1014

[54] Ball J B, Fulmer C B, Larsen J S and Sletten G 1973 Energy levels of ^{94}Ru observed with the ^{96}Ru(p,t) reaction *Nucl. Phys. A* **207** 425–32

[55] Flynn E R, Cizewski J A, Brown R E and Sunier J W 1981 ^{136,138}Ba(t,p) and the systematics of neutron pairing vibrations at $N = 82$ *Phys. Lett. B* **98** 166–8

[56] Mulligan T J, Flynn E R, Hansen O, Casten R F and Sheline R K 1972 (t,p) and (p,t) reactions on even Ce isotopes *Phys. Rev. C* **6** 1802

[57] Ball J B, Auble R L, Rapaport J and Fulmer C B 1969 Levels in ^{140,142,144}Nd and a search for pairing vibrations at $N = 82$ with the (p,t) reaction *Phys. Lett. B* **30** 533–5

[58] Flynn E R, van der Plicht J, Wilhelmy J B, Mann L G, Struble G L and Lanier R G 1983 ^{146}Gd and ^{144}Sm excited by the (p,t) reaction on radioactive targets *Phys. Rev. C* **28** 97

[59] Lauritsen T *et al* 2002 Direct decay from the superdeformed band to the yrast line in ^{152}Dy$_{86}$ *Phys. Rev. Lett.* **88** 042501

[60] Kantele J, Stöffl W, Ussery L E, Decman D J, Henry E A, Hoff R W, Mann L G and Struble G L 1983 Observation of an $E0$ isomeric transition from the ^{238}U shape isomer *Phys. Rev. Lett.* **51** 91

[61] Singh B, Firestone R and Chu S Y F 1997 table of superdeformed nuclear bands and fission isomers (www version) *Technical Report* LBL-38004, LBL

[62] Singh B, Zywina R and Firestone R B 2002 table of superdeformed nuclear bands and fission isomers: 3rd edn (October 2002) *Nucl. Data Sheets* **97** 241–592

Chapter 4

How prevalent is shape coexistence in nuclei? Open-shell and global views

The view of nuclear shape coexistence is extended into open-shell regions. The need for detailed spectroscopic studies and the types of data needed are categorized. A view to global and unified perspectives is explored.

Concepts: comprehensive spectroscopy, energy parabolas.

Learning outcomes: the key data view from this chapter is the exploration of the occurrence of coexisting shapes in nuclei wherein detailed spectroscopic study is essential. In particular, the role of electric monopole transitions is illustrated. Further, the need to consider two-state mixing of coexisting structures is presented. In odd-mass nuclei, key ideas for identifying coexisting structures are introduced: particularly the recurring 'parabolic' energy patterns.

A major challenge to elucidating shape coexistence across the full extent of the Chart of the Nuclides is the level of spectroscopic detail that is necessary. There are many published suggestions of the presence of shape coexistence, based on proxies, but spectroscopic proof is lacking. A key requirement is the need for comprehensive (complete) spectroscopy up to a specific excitation energy in a specific nucleus. In this chapter, we look at exploration of shape coexistence in open-shell nuclei and then we turn to a consideration of a global and unified perspectives.

4.1 Shape coexistence in open-shell nuclei

The power of mapping $E0$ transition strengths is that they reveal shape coexistence that is obscured by mixing. This is illustrated herein for the $N = 90$ isotones, with evidence for a similar occurrence of shape coexistence for the $N = 60$ isotones. Key features of these occurrences are presented in figures 4.1 and 4.2 for the zirconium isotopes at and near $N = 60$, and in figure 4.3 for the $N = 90$ isotones.

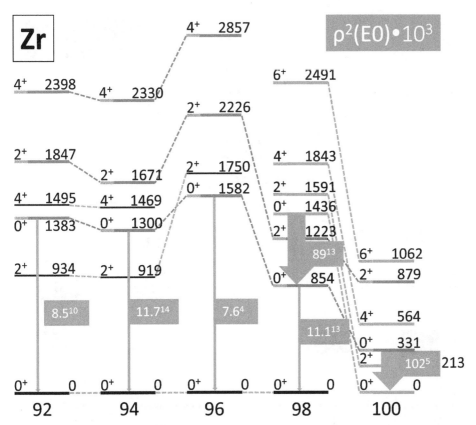

Figure 4.1. Systematics of the even-mass zirconium isotopes between $N = 52$ and $N = 60$: evidence for shape coexistence from $E0$ transition strengths given as $\rho^2(E0) \times 10^3$ values. Deformed band structures are suggested by highlighting levels in blue and orange. Spherical structures appear as black levels. Reprinted from [1], copyright (2022) with permission from Elsevier.

The $N = 60$ region, centred on the zirconium isotopes, was long ago recognized to be unusual in that there was a sudden and massive change in the energies of first-excited 2^+ states between ^{98}Zr and ^{100}Zr [2]. A global view of the significance of changes in ground-state properties at and near $N = 60$ and $N = 90$ is depicted in figure 4.4. The ground-state properties, as revealed by isotope shift data and two-neutron separation energies (deduced from mass measurements), show a sudden increase in mean-square charge radius and in binding energy at $N = 60$ and at $N = 90$, centred on $Z = 39$ and $Z = 63$, respectively. The increase in binding energy can be attributed to the nuclei deforming. A more detailed view of isotope shift data is presented as differences in isotope shifts, so-called Brix–Kopfermann plots, in figure 4.5. This reveals the localization of the most dramatic changes as confined to Nd–Dy and their neighbouring odd-Z isotopes. A more detailed view of the role of $N = 90$ is depicted in figure 4.6, which shows the isotope shift data for the europium isotopes.

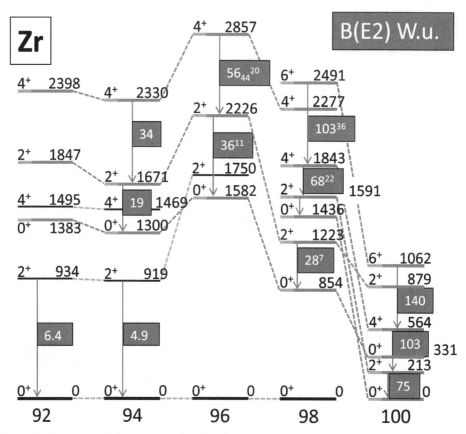

Figure 4.2. Systematics of the even-mass zirconium isotopes between $N = 52$ and $N = 60$: evidence for shape coexistence from $E2$ transition strengths, $B(E2)$ in W.u. The B_{20} values for ^{96,98}Zr are 2.3, 2.9 W.u., respectively. Reprinted from [1], copyright (2022) with permission from Elsevier.

The $N = 90$ region, specifically ^{152}Sm was only more recently shown to involve shape coexistence via a variety of spectroscopic probes. The key motivation for the study was the pattern of $E0$ strengths evident in figure 4.3, which are related to mixing of coexisting bands shown in a schematic manner in figure 4.7. The spectroscopy required detailed quantification of $E2$ transition strengths, especially between the states in the excited $K = 0$ band, shown in figure 4.3. These $E0$ and $E2$ transition strengths then permitted a two-band mixing calculation. The details of the energies used in the calculation, and the resulting description of the $E0$ transition strengths and isomer shift for the 2_1^+ state, are shown in figure 4.8. The degree to which such a calculation fails is presented in figure 4.9. Figure 4.10 shows a feature, unique to the rotor model description of the $N = 90$ isotones: the $E2$ transition strengths in the ground-state bands exceed those of the simple rotor model, cf table 1.1 and figure 1.1. The explanation is evident from the example of ^{152}Sm: it results from shape coexistence and mixing.

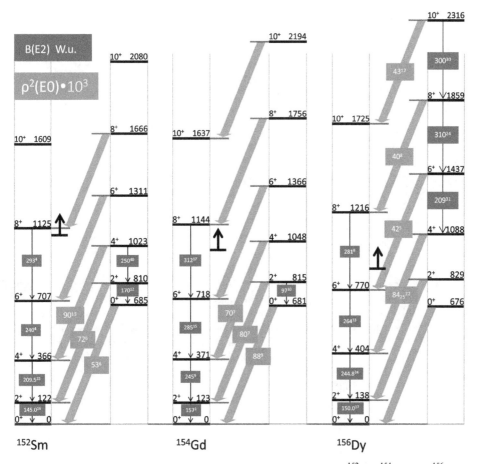

Figure 4.3. *E*0 and selected *E*2 transition strengths in the *N* = 90 isotones, ^{152}Sm, ^{154}Gd, and ^{156}Dy. *E*0 transitions without strength assignments are due to lack of lifetime data. Horizontal bars with upward-pointing arrows indicate energies above which levels with positive parity are omitted. Reprinted from [1], copyright (2022) with permission from Elsevier.

4.2 Global and extended views of shape coexistence

The foregoing details point to patterns that can be investigated by systematic study, directly via *E*2 and *E*0 transition strengths and indirectly via excitation energies and transfer reaction spectroscopy. We have noted that energy patterns are a useful proxy for shape coexistence in nuclei and are far less demanding of spectroscopic technique than transition strengths.

A particularly amenable feature to systematic study of energy patterns related to shape coexistence is that of intruder bands in odd-mass nuclei. Some hints to the patterns of such structures are shown in figures 3.5, 3.7–3.9 and 3.15. The global

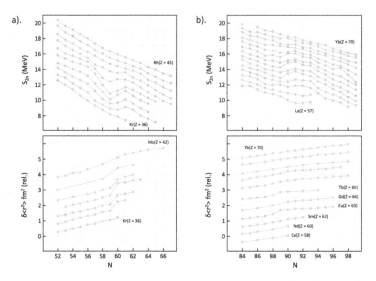

Figure 4.4. Ground-state properties in the region of (a) $N = 60$, $Z = 40$ and (b) $N = 90$, $Z = 62$, as manifested in S_{2n} values and isotope shift data. Note the discontinuities in the systematic trends with increases in S_{2n} values at $N = 60$, 61 and at 90, 91. Increases in $\delta\langle r^2 \rangle$ values at or near these neutron numbers are also evident, but the pattern is less clear. An enhanced view of $\delta\langle r^2 \rangle$ values for the $N = 90$ region is shown in figure 4.5 as differences in $\delta\langle r^2 \rangle$ values and an expanded scale view of the Eu isotopes is presented in figure 4.6. The significance of these ground-state properties as signatures of shape coexistence is discussed in detail in the text. The data are taken from [3] and [4] (with an exception made for the Rb isotopes where there are errors in [4] and the original source is used [5]).

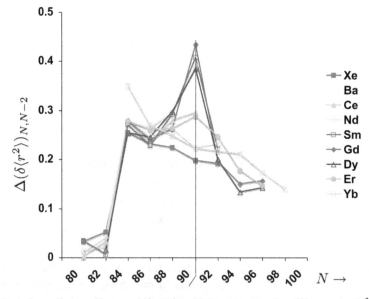

Figure 4.5. An 'enhanced' view of isotope shifts at $N = 90$ near $Z = 64$, using differences in $\delta\langle r^2 \rangle$, i.e. double differences in mean-square charge radii. Portrayal of isotope-shift data in this manner is termed a Brix–Kopfermann plot. The figure is made available courtesy of W D Kulp. The data are taken from [6] and see [4].

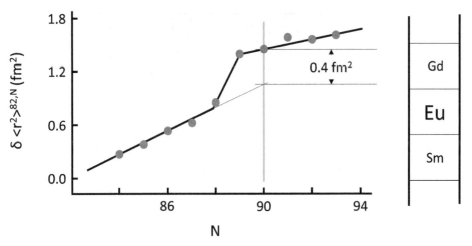

Figure 4.6. A view of isotope shift data for the europium isotopes ($Z = 63$). This provides a measure of the differences in mean-square charge radii associated with a shape coexistence interpretation of structure at $N = 90$, as discussed in the text. The data are taken from [4].

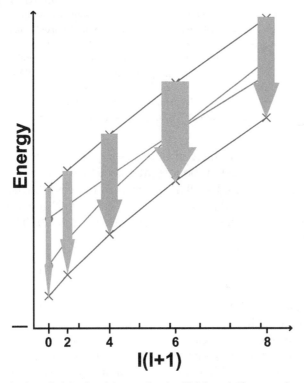

Figure 4.7. A schematic view of mixing involving two bands which 'cross'. Shape coexistence is implicit in such an energy versus $I(I + 1)$ plot. The unmixed bands are shown in red, the mixed bands in black, and the resulting $E0$ transition strength is presented, qualitatively, by the widths of the orange arrows. Note that the resulting mixed structures can appear as identical bands. This is quantified in figure 4.8. Reprinted from [1], copyright (2022) with permission from Elsevier.

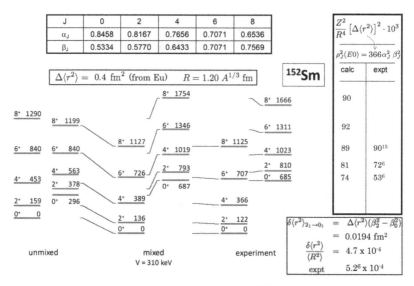

Figure 4.8. A two-band mixing view of the low-energy structure of ^{152}Sm. The $E0$ properties are presented and the $E2$ properties are characterized in detail in appendix B. The isomer shift between the ground state and the 2^+_1 state is also included. The value of $\Delta\langle r^2 \rangle$ is taken from figure 4.6. Note that the mixing strength is constant, independent of the spins of the mixed configurations. As a guide to the unmixed configurations, the ground-state band in ^{148}Ce (an $N = 90$ isotone) is adopted for the less-deformed structure and the ground-state band in ^{154}Sm is adopted for the more-deformed structure. The relative energies of these unmixed configurations is set by making the 6^+ band members degenerate in energy: this results in maximal $E0$ strength for the $6^+ \rightarrow 6^+$ transition, as observed, cf figure 4.7. Note that the $\Delta\langle r^2 \rangle$ value in combination with maximal mixing results in $\rho^2(E0) \times 10^3 = 92$ which can be compared with the observed value, 90 ± 15. This calculation is similar to one given in [7]. Reprinted from [1], copyright (2022) with permission from Elsevier.

energy patterns for intruder states in odd-Z isotopes[1] with respect to $Z = 50$ are shown in figures 4.11–4.13. 'Parabolic' patterns in these energies are evident, with energy minima near to the mid-shell at $N = 66$. But there is a further pattern evident in figure 4.13 which shows that the energy minima 'shift' towards lower neutron number on either side of the closed proton shell at $Z = 50$. There is no current explanation of this.

Intruder bands in odd-Z isotopes adjacent to $Z = 82$ provide a very similar view to that near $Z = 50$. A broader view, cf figures 3.8, 3.9 and 3.15, of the thallium isotopes ($Z = 81$) is presented in figure 4.14. Immediately, one recognizes two sets of states in the lightest thallium isotopes that suggest the emergence of a second parabola. Indeed, a further view of multiple intruder bands in the thallium isotopes is available and is shown in figure 4.15. These structures show different energy minima for the parabolic energy patterns. They also show different spin sequences, which are discussed below. Matching the thallium systematics, there is a similar pattern in the bismuth isotopes ($Z = 83$) shown in figure 4.16. The parabolic energy

---

[1] We note that these patterns were recognized by the late David Fossan in a program of systematic study of the Sb, I and Cs isotopes [8–17].

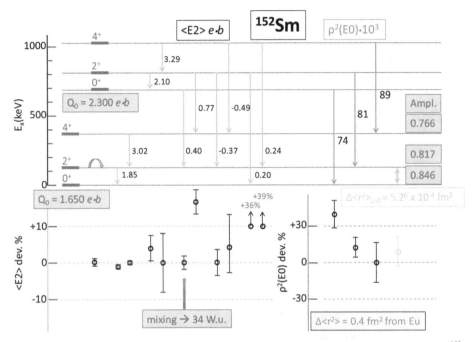

Figure 4.9. Results of the two-band mixing description for $\langle E2 \rangle$ and $\rho^2(E0) \times 10^3$ properties of ^{152}Sm, depicted as deviations from experimental values, (calc. − expt.)/expt. ×100%. The numbers shown are the calculated values. Here, the details are stated in terms of the number of parameters used in the $J = 0, 2, 4$ subspace: three mixing amplitudes and two intrinsic $E2$ matrix elements, with no free parameters for the E0 properties. The major failures are highlighted in red, i.e. at the 35% level. It is important to note that the inter-band transition strength manifested in $B(E2: 0_2^+ \rightarrow 2_1^+) = 33$ W.u., is entirely due to mixing, i.e. there is no intrinsic inter-band strength in the calculation. This is an important caution for interpreting inter-band transition strength as 'intrinsic', e.g. as evidence for collective vibrations such as a so-called beta vibration. Such a degree of freedom is not supported in ^{152}Sm. Reprinted from [1], copyright (2022) with permission from Elsevier.

patterns are especially pronounced in the gold isotopes ($Z = 79$), evident in figure 4.17.

The gold isotopes are the focus of a longstanding and actively continuing investigation because of their accessibility to a variety of detailed spectroscopic probes. Multiple band structures have been identified. An extended view is presented in figure 4.18. The lowest intruder band is the ground-state structure in ^{181,183,185}Au and details are shown in figure 4.19. The elucidation of such details, which involve very low energy transitions, is extremely demanding on spectroscopic technique. Indeed, some level energies are only known from energy differences between transitions from common higher energy levels (see, e.g. [20]). Generally, very low energy transitions are highly converted and so electron spectroscopy is the preferred direct means for observing them; however, low-energy electron spectroscopy requires ultra-thin sources to avoid energy degradation in the radioactive source. For very low energies ($E_e < 15$ keV) it also requires a negative voltage bias on the

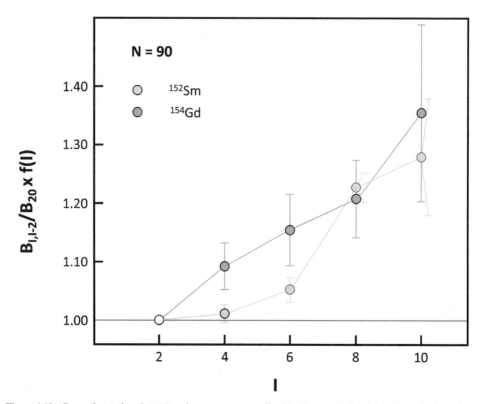

Figure 4.10. Ground-state band $B(E2)$ values, $B_{I,I-2}$ normalized to B_{20} and to the rigid rotor spin dependence $f(I)$, cf equation (1.6). The data points are for the $N = 90$ isotones, ^{152}Sm (blue) and ^{154}Gd (orange). Comparison with 1.1 would naively suggest that these are 'soft' rotors with centrifugal stretching of their intrinsic quadrupole moments; whereas the two-band mixing analysis for ^{152}Sm, cf figures 4.8 and 4.9 and see exercise B-5, reveals that the effect is the result of mixing of coexisting bands. (See exercise 4-1 for a two-band mixing analysis of ^{154}Gd.) The data are taken from ENSDF.

source to 'boost' the electron energy into the dynamic range of the spectrometer (see, e.g. figure 6.12 in [21] and [22]).

The spin sequences in the bands presented above exhibit smooth systematic trends, variously as $I, I + 1, I + 2, \ldots$ (Bi isotopes, $I^\pi = 1/2^+$; heavy Tl isotopes, $I^\pi = 9/2^-$; ^{187,189}Tl, $I^\pi = 13/2^+$) and as $I, I + 2, I + 4, \ldots$ (light Tl isotopes, $I^\pi = 9/2^-$, $13/2^+$; Au isotopes, $I^\pi = 9/2^-$, $7/2^-$, $13/2^+$). The $B(E2)$ values manifested in the $I, I + 1, I + 2, \ldots$ sequences are weak and those in the $I, I + 2, I + 4, \ldots$ sequences are strong. Some remarks on spin sequences for intruder bands have already been made pertaining to prolate versus oblate deformation and $K = 1/2$ versus $K = j$ Nilsson states: notably that intruder bands below a closed shell are either prolate with $K = 1/2$ (In, Ag, Rh) or oblate with $K = j$ (Tl). A characteristic feature of $K = j$ bands is that they have the spin sequence $I, I + 1, I + 2, \ldots$ and of $K = 1/2$ bands is that they are 'decoupled' with a displacement of the $I, I + 2, I + 4, \ldots$ spin sequence relative to the $I + 1, I + 3, \ldots$ spin sequence. In

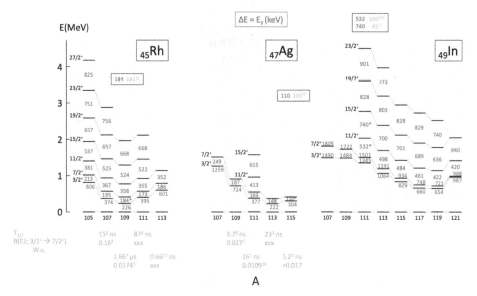

Figure 4.11. Systematic features of intruder states in the odd-mass Rh, Ag and In isotopes. The states shown are the favoured states from the signature-split members of the rotational bands built on the $1/2^+$ [431] Nilsson configuration, which stems from the $\pi 1g_{7/2}$ spherical shell model configuration. The half lives and the hindered electromagnetic decay strengths of the band heads, where known, are given in green at the bottom of the figure. Transition energies are shown in red and some level energies are shown in blue. A few intra-band transitions have known (enhanced) electromagnetic decay strengths and these are given in boxes over the bands, with transition energies in red and $B(E2)$ values, in W.u., in green; the values in ^{111}In (marked with asterisks on the figure) would appear to need remeasurement. The data are taken from ENSDF.

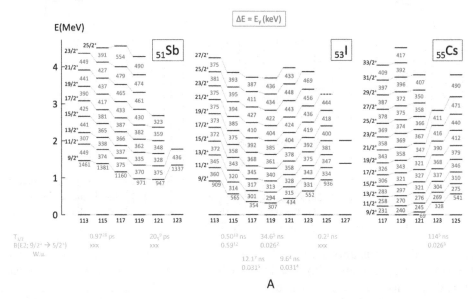

Figure 4.12. Systematic features of intruder states in the odd-mass Sb, I and Cs isotopes. The states shown are members of the rotational bands built on the $9/2^+$ [404] Nilsson configuration, which stems from the $\pi 1g_{9/2}$ spherical shell model configuration. Transition energies are given in red and band-head energies are given in blue. The half lives and the hindered electromagnetic decay strengths of the band heads, where known, are given in green at the bottom of the figure. The data are taken from ENSDF.

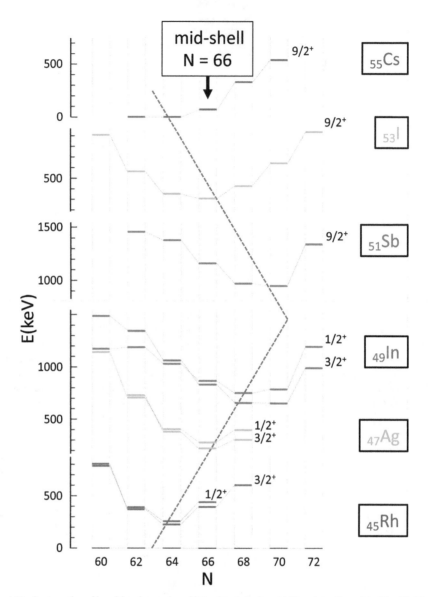

Figure 4.13. Systematics of band-head energies of intruder states in odd-Z nuclei adjacent to $Z = 50$. Note the 'parabolic' trends in these energies and the systematic 'shift' to lower neutron numbers both above and below $Z = 50$. The diagonal lines are to guide the eye to these shifts. For the $1/2^+$ [431] bands, the lowest $1/2^+$ and $3/2^+$ band members are shown. The data are taken from ENSDF.

turn, intruder bands above a closed shell can be expected to be either prolate with $K = j$ (Sb, I, Cs) or oblate with $K = 1/2$. The intruder bands in the Bi isotopes appear to be dominated by $j = 1/2$, cf figure 3.15 (but note [191]Bi in figure 4.16). However, there are no established rules for the appearance of oblate versus prolate deformation in nuclei. The only observation that can be made is that prolate deformation predominates in nuclei: there is no consensus as to why this occurs.

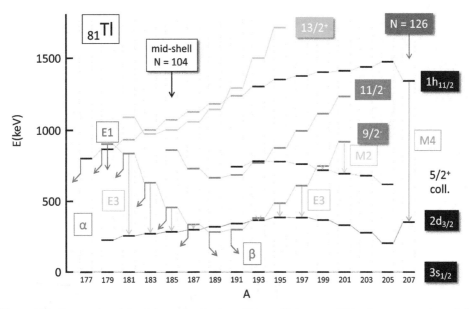

Figure 4.14. Systematics of low-lying states in the odd-mass Tl isotopes. This gives an overview of the manifestation of intruder states in these isotopes, notably stemming from the $\pi 1h_{9/2}$ and $\pi 1i_{13/2}$ spherical shell model configurations. The first band member associated with the $1h_{9/2}$ configuration, i.e. an oblate $9/2^-$ [505] band, is shown. A rich variety of decay modes are shown in association with the $9/2^-$ band heads. Note that the $1h_{11/2}$ state has not yet been identified in $^{181-191}$Tl, the annotation is to suggest its excitation in these isotopes; but this state has been identified in ^{177,179}Tl. See [18] for details attributed to the $9/2^-$ isomer in ^{179}Tl. Other data are taken from ENSDF.

The view presented above leads to a unifying perspective for odd and even nuclei via the energy parabolas. The leading example is presented in figure 4.20. This systematic view supports the interpretation of intruder state structures as dominated by proton pairs (particle pairs, hole pairs, or both) correlated with the many neutrons across the open shell. This naturally leads to a maximum correlation energy at the mid-shell for the neutrons, i.e. a minimum in the parabolas at $N = 104$ for $Z \sim 82$. Indeed, figure 4.21 confirms this perspective for $Z \sim 50$. (We view the multiple parabolas with shifts in the energy minima as a 'higher-order effect', yet to be explained.) Note that these energy trends are independent of whether the bands are strongly coupled or decoupled and, by implication, independent of whether they are prolate or oblate.

With the global perspective manifested in figures 4.20 and 4.21, and the more detailed patterns that have led to this view, we can enquire to what degree such patterns are manifested at and near other shell closures. A view is provided in figure 4.22 for the $Z = 20$ closed shell. Indeed, the even-mass Ca isotopes exhibit proton-pair excited states at low energy, albeit coexisting with multiple-pair excited states, as depicted in figure 4.23. The spectroscopic evidence for the multi-pair excitations is summarized in table 4.1. An important aspect of these assignments is the 'multi-hole' configuration of each target nucleus in the multi-nucleon transfer

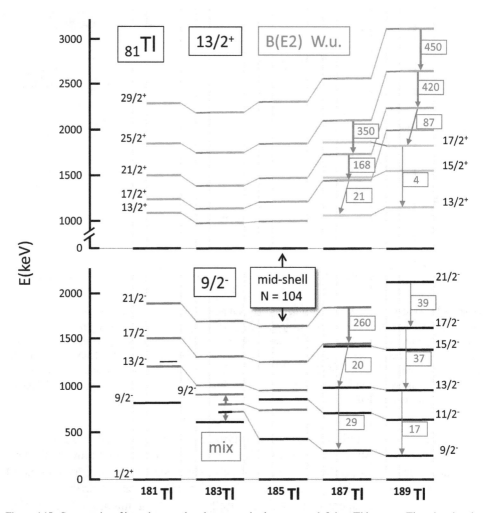

Figure 4.15. Systematics of intruder state band structure in the neutron-deficient Tl isotopes. These involve the Nilsson configurations $9/2^-$ [505], $13/2^+$ [606] and the highly deformed Nilsson configurations assigned as $1/2^-$ [541] and $1/2^+$ [660]. The evidence for the shape coexistence resides in the $B(E2)$ values, shown in W.u.; and is implicit in the decoupled and strongly coupled spin sequences, which is discussed further in the text. The energy pattern in ^{183}Tl suggests a mixing and repulsion of the $9/2^-$ members of the [505] and [541] bands. The data are taken from ENSDF.

reaction. Already ^{42}Ca is a nucleus of focal interest, cf figures 3.20 and 3.21. We address other closed shells below.

Figure 3.28 summarizes what is known about pair excitations in even–even nuclei with $N = 50$ and 82. A parabolic energy pattern is lacking. However, there are useful hints to the fate of the parabolic pattern at $N = 50$ and $N = 82$ from intruder states in the $N = 49$ and $N = 81$ isotones, shown in figures 4.24 and 4.25, respectively. We have encountered the effect of the $Z = 40$ configuration in 'suppressing' the

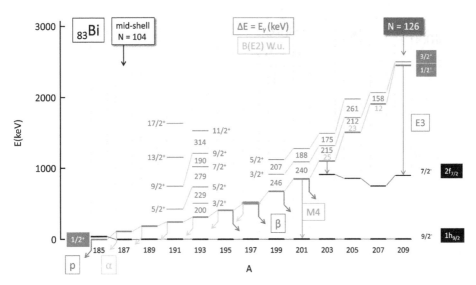

Figure 4.16. Systematics of low-lying states in the odd-mass Bi isotopes. This gives an overview of the manifestation of intruder states in these isotopes, notably stemming from the $\pi 3s_{1/2}$ spherical shell model configuration. Transition energies in keV are shown in red and $B(E2)$ values in W.u. are shown in green. A rich variety of decay modes are shown in association with the $1/2^+$ band heads. There appears to be a change from a strongly coupled band in ^{193}Bi to a decoupled band in ^{191}Bi. The data are taken from ENSDF; and the isomerism in ^{185}Bi is established in [19].

emergence of deformation in nuclei between $N = 50$ and $N = 58$, centred on the zirconium isotopes, cf figures 4.1 and 4.2. We make the interpretation that this also results in a suppression of the parabolic minimum for intruder states in $N = 50$ isotones at $Z = 40$. This is schematically illustrated in figure 4.26. A subshell suppression can be invoked to explain the pattern of shape coexistence at $N = 20$, implied in 3.17, and shown with respect to $N = 19$ and 21 intruder states in figure 4.27. The implication is that $Z = 14$ 'suppresses' the parabolic energy trend. An in-depth look at the systematics of the $N = 19$ and 21 isotones is presented in [34, 35].

The manifestation of shape coexistence in the zirconium isotopes, evident in figure 4.2, suggests a repeat of the parabolic pattern with respect to the interpretation of $Z = 40$ as a closed subshell combined with $N = 56$ acting to further suppress collective structures at low energy. This is depicted schematically in figure 4.28. The similarity between the suppression of collective structures in the region of $Z = 40$, $N = 60$ and $Z = 64$, $N = 90$ can be viewed from this perspective. This is a perspective that needs a deeper exploration.

We extend this view with the information on intruder states at $N = 28$, shown in figure 4.29. A more speculative view of hints at intruder states in the vicinity of the nickel isotopes, $Z = 28$, is shown in figure 4.30, which support intruder states in the neutron-rich copper isotopes, $Z = 29$. Thus, the neutron-rich nickel isotopes may be expected to exhibit low-energy shape coexistence. This has been suggested in

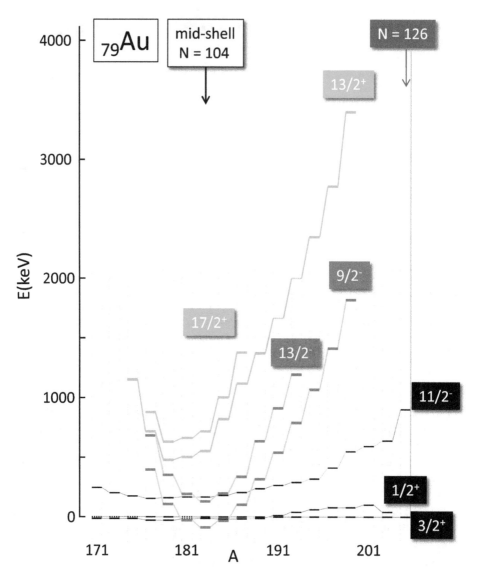

Figure 4.17. Systematics of intruder state structures and selected low-energy states in the odd-mass Au isotopes. The intruder states stem from the $\pi 1h_{9/2}$ and $\pi 1i_{13/2}$ spherical shell model configurations. The selected low-energy states stem from the $\pi 3s_{1/2}$, $\pi 2d_{3/2}$ and $\pi 1h_{11/2}$ spherical shell model configurations. The data are taken from ENSDF and from sources given in figure 4.19.

^{70}Ni [38]. Suppression of collectivity at mid-shell can be interpreted as due to a subshell effect at $N = 40$.

Energy parabolas provide some perspective on the occurrence of shape coexistence in nuclei further removed from closed shells. For example, figure 4.31 shows the intruder state patterns in the odd-Z nuclei: Ir ($Z = 77$), Re ($Z = 75$), $\cdots$Ho ($Z = 67$)—note the energy minima exhibit the trend towards lower neutron numbers with increasing distance from the $Z = 82$ shell closure, as observed for $Z = 50$ in

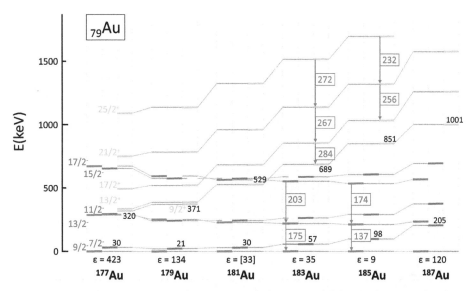

Figure 4.18. Systematics of intruder state band structure in the neutron-deficient Au isotopes. These involve the intruder state configurations stemming from the $\pi 1h_{9/2}$, $\pi 2f_{7/2}$ and $\pi 1i_{13/2}$ spherical shell model configurations. The evidence for strong deformation resides in the $B(E2)$ values, shown in W.u. The data are taken from [23–25] and ENSDF.

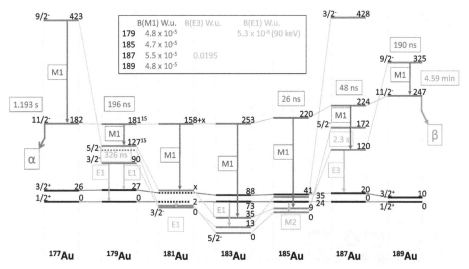

Figure 4.19. Detailed view of the lowest energy intruder states (shown in red) in the neutron-deficient Au isotopes. These involve the intruder state configurations stemming from the $\pi 1h_{9/2}$ and $\pi 2f_{7/2}$ spherical shell model configurations and match the $1/2^-$ [541] and $1/2^-$ [530] Nilsson configurations (which are strongly decoupled). The weakly deformed states associated with the $\pi 3s_{1/2}$, $\pi 2d_{3/2}$ and $\pi 1h_{11/2}$ structures are also shown. The highly retarded M1 transitions between the $9/2^-$ intruder states and the $11/2^-$ states associated with the $1h_{11/2}$ structure are presented, together with parity-changing transitions and other decay modes. Data are taken from [20, 23, 26–28] and ENSDF.

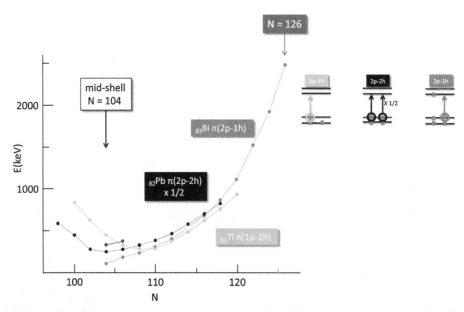

Figure 4.20. A summary view of shape coexisting/intruder structures for $Z = 82$, showing the relationship between the $9/2^-$ intruder states in the Tl isotopes, the $1/2^+$ intruder states in the Bi isotopes and the candidate deformed 0^+ states in the Pb isotopes. These structures can be viewed from their dominant particle–hole character, viz. Tl $\pi(1p–2h)$ states, Bi $\pi(2p–1h)$ states and Pb $\pi(2p–2h)$ states: thus, it is evident that the pair structures, particle or hole, dominate equally. The Pb states, involving two pairs are therefore shown with excitation energies divided by two. The structures below $N = 108$ indicate multiple coexisting structures, which lies beyond the present focus.

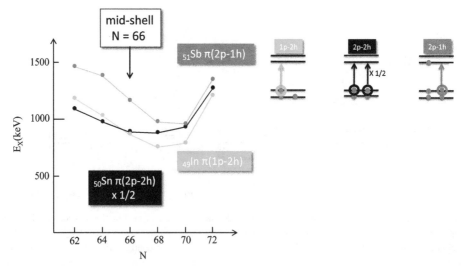

Figure 4.21. A summary view of shape coexisting/intruder structures for $Z = 50$, showing the relationship between the $1/2^+$ intruder states in the In isotopes, the $9/2^+$ intruder states in the Sb isotopes and the candidate deformed 0^+ states in the Sn isotopes. The pattern resembles that for $Z = 50$, depicted in figure 4.20.

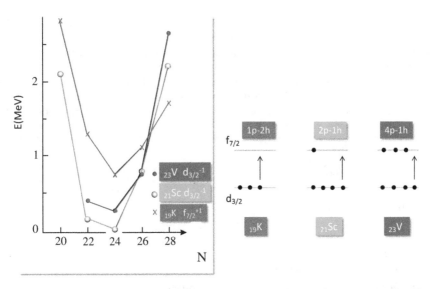

Figure 4.22. Intruder states in the $Z = 19$ (potassium), $Z = 21$ (scandium) and $Z = 23$ (vanadium) isotopes. The 'parabolas' are narrow in this region because they are confined by the $N = 20$ and $N = 28$ shell closures. It is not possible to make a direct comparison with deformed states in the even-mass calcium isotopes because they involve multi-particle–multi-hole excitations: these are summarized in figure 4.23.

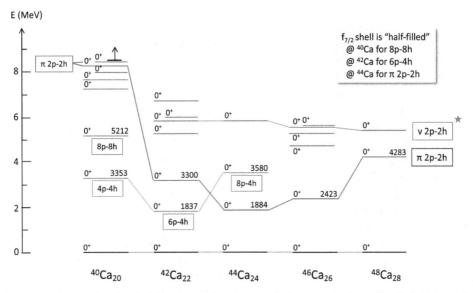

Figure 4.23. Systematics of excited 0^+ states in the even-mass Ca isotopes with $20 \leqslant N \leqslant 28$. Multi-particle–multi-hole configurations are indicated, where there is evidence of dominance from multi-nucleon transfer reaction spectroscopy (the red star indicates neutron-pair excitation w.r.t. $N = 28$): see table 4.1 for more details. The figure is adapted from [29], copyright (1992), with permission from Elsevier.

Table 4.1. Spectroscopic evidence for assignment of multi-particle-multi-hole configurations to excited 0^+ states in ^{42,44,46}Ca from multi-nucleon transfer reactions. A critical aspect of these transfer reactions is the 'hole' and 'particle' configurations that can be assigned to the target nuclei. Note, these multi-particle-multi-hole assignments are only important components to what likely involve multi-shell configurations involving particles in shells beyond those immediately above the Fermi energy. The structures are indicated as π(2p–2h) (violet), ν(2p–2h) (red) and 4p–4h $\times$ ν1f$_{7/2}^{n}$ (blue), where 'p' = particle and 'h' = hole. The designations 'strong' and 'weak' imply that the 0^+ states are strongly or weakly populated in the transfer reactions. Energies are given in MeV. The references from where the data derive are given in the last column.

	0_1^+	0_2^+	0_3^+	0_4^+	0_5^+	0_6^+	References
^{42}Ca	0.00	1.84	3.30	5.35	5.86	6.02	
^{40}Ca(t,p)^{42}Ca					Strong		[30]
^{40}Ar(^{3}He,n)^{42}Ca		Weak					[31]
^{38}Ar(^{6}Li,d)^{42}Ca			Strong				[32]
^{44}Ca	0.00	1.88	3.58	5.86			
^{48}Ti(d,^{6}Li)^{44}Ca			Strong				[31]
^{46}Ti(^{14}C,^{16}O)^{44}Ca		Strong					[33]
^{42}Ca(t,p)^{44}Ca				Strong			[30]
^{46}Ca	0.00	2.42	4.76	5.32	5.60	5.63	
^{48}Ti(^{14}C,^{16}O)^{46}Ca		Strong					[33]
^{42}Ca(t,p)^{44}Ca					Strong	Strong	[30]

figure 4.13. This suggests that shape coexistence should appear at low energy in even–even nuclei adjacent to these minima. An extended realization of this expectation is present in the platinum isotopes centred on $N = 104$. We particularly note that the indicator manifested in figure 4.31 is persuasive because intruder configurations in odd-mass nuclei do not (strongly) mix with other configurations: this is due to their parities. However, intruder configurations in even–even nuclei, due to their paired structure, can mix with other configurations. This is implied for the Pt isotopes in figure 4.32, which was the initial clue to the interpretation of the ground states of these isotopes being dominated by intruder configurations. Some details of the Pt isotopes follow.

Figure 4.33 provides key details for the systematics of the even-mass platinum isotopes. There is the suggestion of intruding structure with a parabolic pattern, centred on the mid-shell point at $N = 104$. The $B(E2)$ values for the lowest transitions leading to the ground states are shown and they reveal a large increase for the isotopes that would be candidate nuclei for possessing intruder states as ground states. However, isotope shift data, shown in figure 4.34 show only a moderate increase, with the implication that the intruder configurations must be mixing with the valence shell configurations.

The global view provided by figure 4.34 points to many avenues of exploration. Notably, in the open shell above $Z = 82$, specifically in the polonium isotopes, shape coexistence should occur. Figure 4.35 shows the systematics of the lowest states in the even-mass polonium isotopes and indeed there is evidence for a pattern that

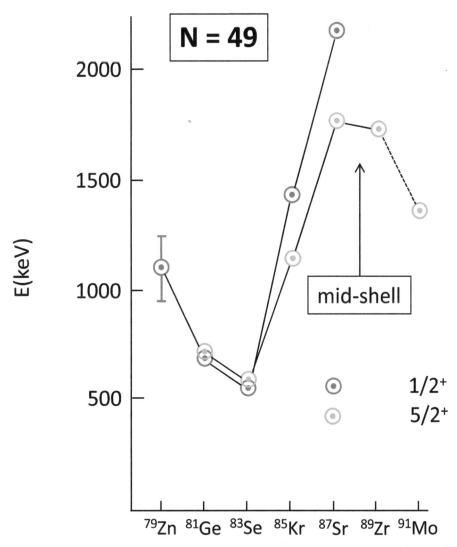

Figure 4.24. Intruder states in the $N = 49$ isotones. These closely match those observed in the $Z = 49$, In isotopes and can be interpreted as dominated by the Nilsson $1/2^+$ [431] configuration stemming from the $\nu 1g_{7/2}$ spherical shell model configuration. Reprinted from [36] under CC-BY-4.0 license.

resembles the platinum isotopes, albeit far less extensive. This is a mass region that is extremely difficult to study because it lies far off the line of stability.

Mixing of intruder configurations with valence shell configurations in even-mass nuclei appears to be universal and may be increasing in nuclei progressively more removed from closed shells, i.e., increasing from mercury, $Z = 80$ to platinum, $Z = 78$, and so on. Thus, $E0$ transitions are expected to feature strongly. This is shown for ^{184}Pt in figure 4.36. At present there are no lifetime data for the states involved and so $E0$ transition strengths are not available.

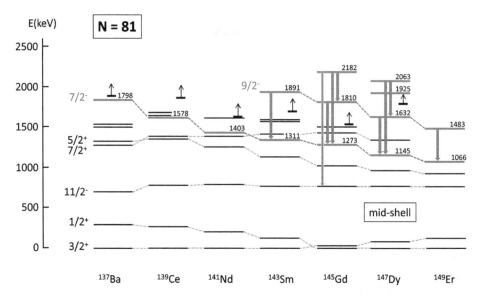

Figure 4.25. Intruder state candidates in the $N = 81$ isotones. These can be interpreted as oblate-deformed bands built on the Nilsson configuration $7/2^-$ [503]. Note that this bears a resemblance to the oblate $9/2^-$ [505] bands observed in the $Z = 81$, Tl isotopes. Note the preference for candidate band members in ^{143}Sm, ^{145}Gd and ^{147}Dy to decay to the $7/2^-$ state, not the $11/2^-$ state (which is a $1h_{11/2}$ shell model state). Reprinted from [36] under CC-BY-4.0 license.

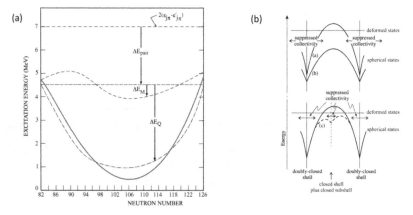

Figure 4.26. Schematic views of intruder configuration parabolas. In (a) the pattern observed when there are no subshell effects is depicted, with a semi-quantitative depiction of the three dominant energy factors due to pairing correlations, ΔE_{pair}, deformation or 'quadrupole' correlations, ΔE_Q and the monopole shift, ΔE_M. In (b) the pattern is inverted to suggest that one can view the intruder configuration parabolas rather as the (rare) intrusion of spherical structures at closed shells. This view is augmented by the pattern that occurs when a subshell is present. Reprinted figures with permission from [37]. Copyright 2011 by the American Physical Society.

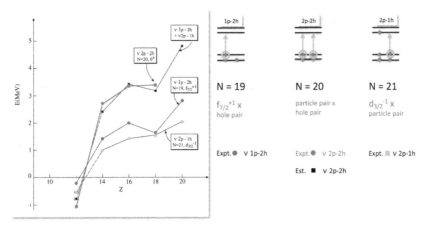

Figure 4.27. A summary view of shape coexisting/intruder structures for $N = 20$, showing the relationship between the intruder states in the $N = 19$ and $N = 21$ isotones and the candidate deformed 0^+ states in the $N = 20$ isotones.

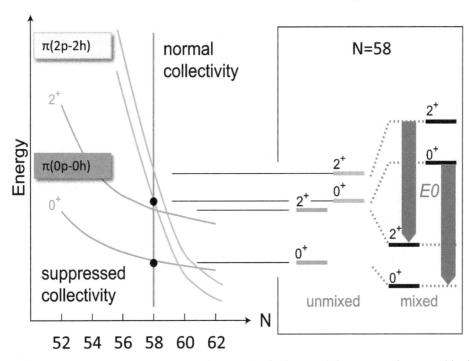

Figure 4.28. A schematic view of the zirconium isotopes showing how coexisting structures invert, resulting in 'sudden' changes in ground-state structure. The coexisting structures are labelled as originating from proton pair structure with respect to $Z = 40$. However, the implication of three coexisting structures in ^{98}Zr (cf figure 4.2) requires a deeper look at the collective structures in this mass region; a discussion of this is beyond the scope of the present work.

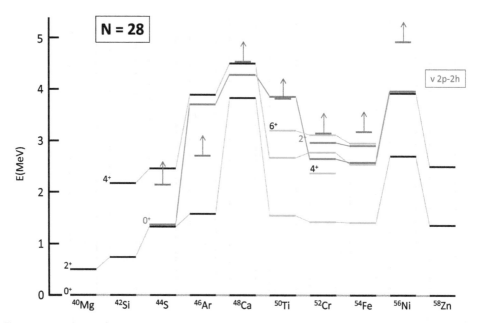

Figure 4.29. Evidence for ν (2p–2h) states in the $N = 28$ isotones. These structures are identified by two-neutron transfer reactions—the t,p reactions: [30, 39–42] and the p,t reactions: [43–45]. The spectroscopy of the 0^+ state in ^{44}S has been studied by two- and four-proton knockout reactions: [46–48]. The seniority-dominated structures in the $1f_{7/2}$ shell nuclei are highlighted in green. Other data are taken from ENSDF.

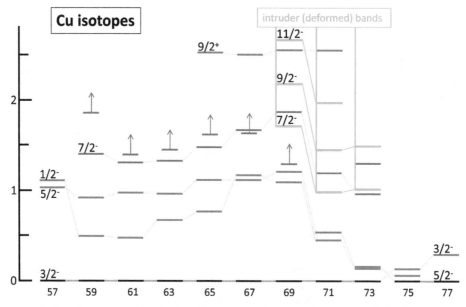

Figure 4.30. Intruder state candidates in the odd-mass copper isotopes. The spectroscopic detail in the heaviest isotopes is incomplete. Color coding of levels is simply to guide the eye.

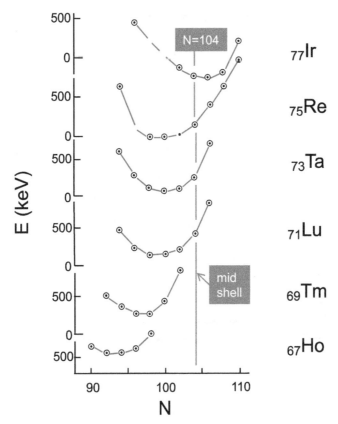

Figure 4.31. Systematics of the lowest-energy intruder configuration in the odd-Z rare earth isotopes. This is the Nilsson configuration $\Omega^\pi[Nnz\Lambda] = 1/2^-[541]$. Note that the minima in the 'parabolas' progressively move to lower neutron numbers with decreasing proton number, cf figure 4.13.

A notable new feature, evident in ^{184}Pt, cf figure 4.36 is the appearance of $E0$ transitions between pairs of excited bands, specifically between a pair of $K = 2$ bands. Such $E0$ decay patterns are seen also in other mass regions and examples are shown for ^{114}Cd in figure 4.37 and for three $N = 90$ isotones in figure 4.38.

A few further indicators to the presence of shape coexistence are worthy of note. Figure 4.39 shows an extended view of figure 4.32 for ^{183}Pt cf ^{182,184}Pt and for ^{75}Kr cf ^{74,76}Kr and ^{79}Sr cf ^{78,80}Sr. This is again interpreted in terms of mixing occurring for the lowest spin structures in even–even nuclei and the suppression of mixing in neighbouring odd-mass nuclei due to spin-parity restrictions when an unpaired nucleon dominates low-energy structure. Indeed, figure 4.39 implies that $E0$ transitions should be present at low-energy in the even-mass krypton isotopes, and this is observed, as shown in figure 4.40. Exploration of $E0$ transitions in the even-mass strontium isotopes remains to be pursued.

Detailed spectroscopic studies are now beginning to provide a systematic view of deformed structures which reveal that they smoothly extend from regions of strong ground-state deformation into regions of weak ground-state deformation, even

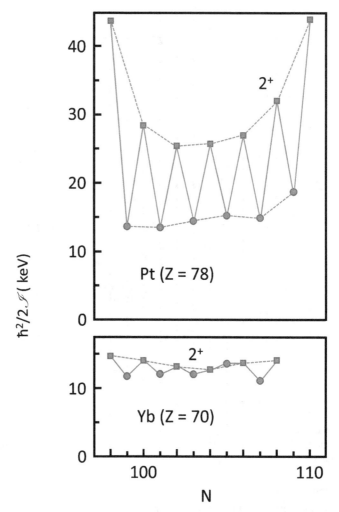

Figure 4.32. Rotational energy parameters, deduced from transition energies, for even- and odd-mass platinum isotopes and for even- and odd-mass ytterbium isotopes; where the ytterbium isotopes are typical for strongly deformed nuclei. The figure is an up-dated view of one appearing in [49].

regions with spherical ground states. Figure 4.41 shows the systematic features of bands built on $K^\pi = 8^-$ isomers in the $N = 106$ isotones. Figure 4.42 shows systematic features of bands built on the Nilsson configuration $7/2^-$ [514] in the $N = 105$ isotones. Note that the $K^\pi = 8^-$ configuration is dominated by the neutron broken-pair configuration $7/2^-$ [514] + $9/2^+$ [624]. This view supports the perspective presented in figure 4.26(b), i.e., that strong deformation exists for all nuclei with $N = 105$ and 106; but in the nuclei with $Z > 76$, weakly deformed (spherical) states come low in energy, even becoming the ground state at and near $Z = 82$. Thus, one may usefully view the spherical configurations as the intruder structures instead of the conventional view of the deformed states being the intruder structures.

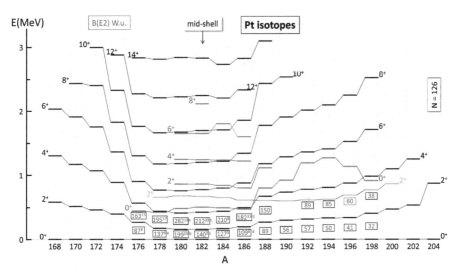

Figure 4.33. Systematics of excited states in the even-mass platinum ($Z = 78$) isotopes. Second excited 2^+ states are shown in green and third excited 2^+ states are shown in red. Second excited 0^+, 4^+, 6^+, $\cdots$ states are shown in red. Selected $B(E2)$ values are shown in blue (in black boxes) with values taken from ENSDF and from 178: a—[50]; 180: b—[51]. cf [52] $B_{42} = 193^{16}$ W.u.; 182: c—[53]; 186: d—[54]. The letter annotation identifies the $B(E2)$ values presented in the figure.

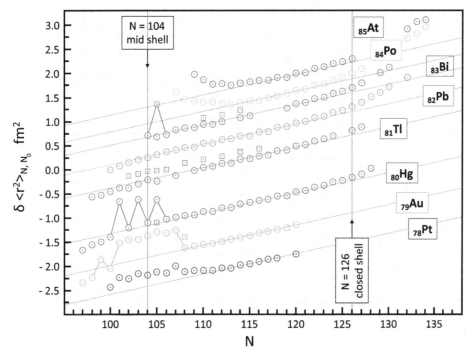

Figure 4.34. Isotope shifts for Pt-At ($Z = 78 - 85$) isotopes. The isotopic sequences are displaced vertically for clarity in showing the systematic trends. Each isotopic sequence possesses a reference isotope with neutron number N_0: this is not defined here as it is irrelevant to the structural message. Note the zig-zag feature in the Hg isotopes, cf figures 3.10 and 3.11. Further note the 'mesa' feature for $^{180-186}$Au, cf figure 4.19. The circle symbols denote ground states and the square symbols denote isomeric (excited) states. The data are taken from [4, 55–68]

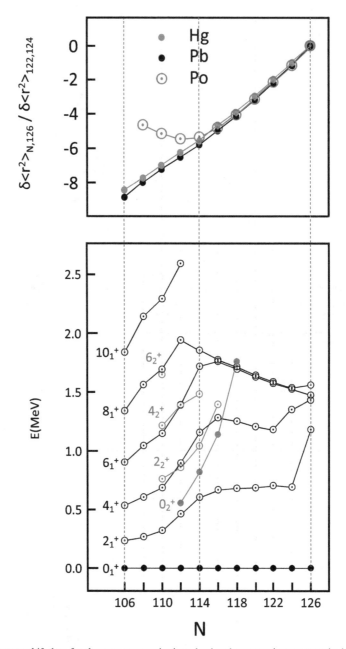

Figure 4.35. Isotope shift data for the even-mass polonium, lead and mercury isotope matched against selected excited states in the even-mass polonium isotopes, as a function of neutron number. Note the sudden increase in mean-square charge radius in the polonium isotopes as the neutron number decreases from $N = 114$ to 112: this suggests that a deformed structure has intruded to become the ground-state structure below $N = 114$. The figure is based on one appearing in [69] using data taken from ENSDF and scaled isotope shift data taken from [67].

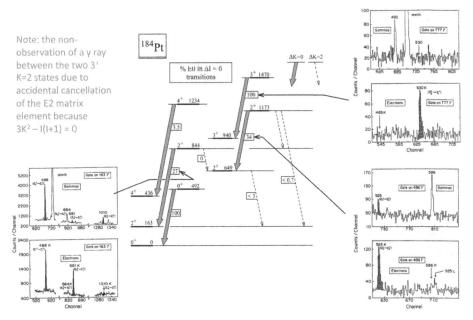

Figure 4.36. The $E0$ decay pattern supporting coexisting bands in ^{184}Pt. Note that $E0$ transitions are not observed between other pairs of spin-2 and spin-4 states, e.g. the 163 and 649 keV pair of states. The evidence for this is based on a precision determination of $E2/M1$-mixing ratios by low-temperature nuclear orientation together with the coincidence-gated conversion electron spectra and corresponding gamma-ray spectra as shown. Reprinted figures with permission from [70]. Copyright 1992 by the American Physical Society.

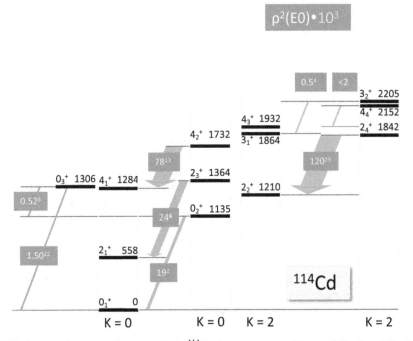

Figure 4.37. Pattern of $E0$ transition strengths in ^{114}Cd that support coexistence of $K = 0$ and $K = 2$ bands. The figure is reproduced from [1], copyright (2022).

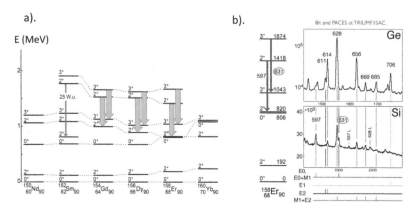

Figure 4.38. (a) $E0$ transitions between $K = 2$ bands in the $N = 90$ isotones. In ^{152}Sm, $E2$ transition strengths establish that the second $K = 2$ band is associated with the second $K = 0$ band. (b) Gamma-ray (Ge = germanium detector) and conversion-electron (Si = lithium-drifted silicon detector) spectra for the beta decay of ^{158}Tm to ^{158}Er (note absence of 631 keV γ-ray line in upper part of (b), cf a similar structure in ^{184}Pt, shown in figure 4.36). The multipolarities of the transitions are indicated as a guide to relative intensities of the conversion-electron peaks versus the gamma-ray peaks. The figure is reproduced from [1], copyright (2022).

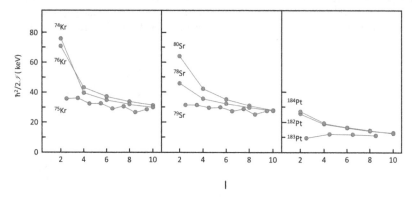

Figure 4.39. Examples of patterns of rotational energy parameters in odd-mass nuclei compared to adjacent even–even nuclei for ^{74,75,76}Kr and ^{78,79,80}Sr, where there is 'convergence' towards a common value. This is similar to the Pt isotopes; which are presented in detail above and are shown here for ^{182,183,184}Pt. The bands in ^{75}Kr and ^{79}Sr have $K = 3/2$ and negative parity and match the Nilsson configuration [301]; the band in ^{183}Pt has $K = 1/2$ and negative parity and matches the Nilsson configuration [521] (only the favored spin states are shown for the plot using $\Delta I = 2$ transitions). The data are taken from ENSDF.

We close with one more perspective on ways to explore manifestations of shape coexistence in even-mass nuclei. This is through the association of intruder states in neighbouring odd-mass nuclei, as depicted in figure 4.43. Some of the low-energy excited 0^+ states shown in figure 4.43 have been identified as involving shape coexistence, the others are yet to be investigated.

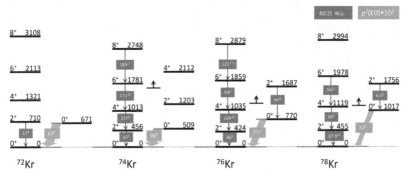

Figure 4.40. $E0$ and selected $E2$ transition strengths in the light krypton isotopes. Reprinted from [1], copyright (2022) with permission from Elsevier.

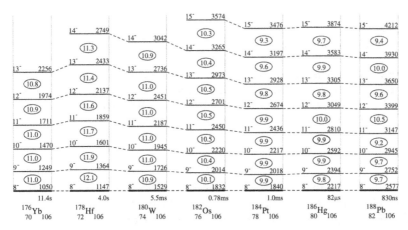

Figure 4.41. Systematics of rotational bands built on the $K^{\pi} = 8^-$ isomeric configuration in the $N = 106$ isotones. This structure is dominated by the broken neutron pair configuration $7/2^-$ [514] $+ 9/2^+$ [624]. Note that the appearance of decreased deformation in ^{178}Hf is due to mixing with the broken proton pair configuration $9/2^-$ [514] $+ 7/2^+$ [404]. The numbers in circles are rotational energy parameters, A (keV), deduced using $E = AI(I + 1)$. Level energies are relative to the ground state in each nucleus. Reprinted figure with permission from [37]. Copyright 2011 by the American Physical Society.

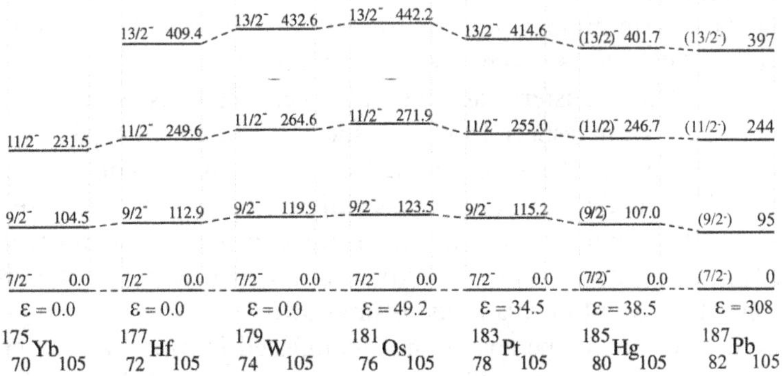

Figure 4.42. Systematics of rotational bands built on the $7/2^-$ [514] Nilsson configuration in the $N = 105$ isotones. The band-head excitation energy, ϵ (keV), is given in each nucleus. The data are taken from ENSDF and [71].

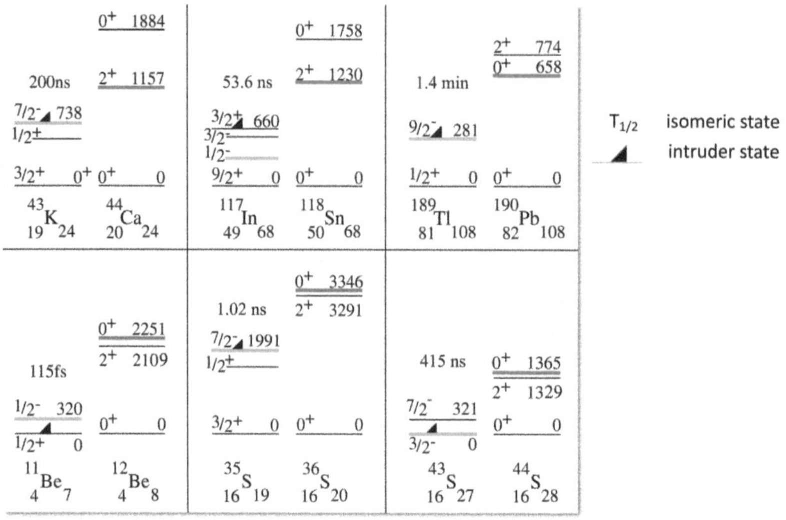

0^+ 1884

200ns 2^+ 1157

$7/2^-$ 738
$1/2^+$

$3/2^+$ 0 0^+ 0

$^{43}_{19}K_{24}$ $^{44}_{20}Ca_{24}$

0^+ 1758

53.6 ns 2^+ 1230

$3/2^+$ 660
$3/2^-$
$1/2^-$

$9/2^+$ 0 0^+ 0

$^{117}_{49}In_{68}$ $^{118}_{50}Sn_{68}$

2^+ 774
0^+ 658

1.4 min $9/2^-$ 281

$1/2^+$ 0 0^+ 0

$^{189}_{81}Tl_{108}$ $^{190}_{82}Pb_{108}$

$T_{1/2}$ isomeric state
◢ intruder state

0^+ 3346
1.02 ns 2^+ 3291

0^+ 2251 $7/2^-$ 1991
2^+ 2109 $1/2^+$

115fs

$1/2^-$ 320 0^+ 0 $3/2^+$ 0 0^+ 0
$1/2^+$ 0

$^{11}_{4}Be_7$ $^{12}_{4}Be_8$ $^{35}_{16}S_{19}$ $^{36}_{16}S_{20}$

415 ns 0^+ 1365
2^+ 1329

$7/2^-$ 321 0^+ 0
$3/2^-$ 0

$^{43}_{16}S_{27}$ $^{44}_{16}S_{28}$

Figure 4.43. A depiction of the occurrence of intruder state isomers adjacent to even–even nuclei possessing low-energy excited 0^+ states. For example, in ^{43}K the ground state is the expected proton hole configuration $1d_{3/2}^{-1}$ and the $7/2^-$ state at 738 keV is the proton 1p–2h $1f_{7/2}^{+1}$ configuration; the lowest excited 0^+ state in ^{44}Ca is at 1884 keV, cf table 4.1 and figure 4.23.

As a summary statement, we note that the term 'intruder' state is commonly used. When an intruder state becomes the ground state this has been termed an 'island of inversion'. One can see from the perspective of the parabolic energy systematics that such occurrences are not islands; but the term will likely stay with us. The term is misleading; and it has misled people to think that the manifestation of shape coexistence in nuclei forms 'islands' that are isolated mass regions defined by sudden changes in nuclear self-organization in their ground states. Nuclei self-organize into a variety of deformations, and these deformations *coexist*. The ordering of these different deformations in a nucleus is simply different in different nuclei; but these orderings have systematic, well-defined patterns. We conjecture that 'coexistence is everywhere' in nuclei.

A further issue of terminology is the use of 'particle–hole' language to describe intruder states. This stems from the preferential population of intruder state candidates by nucleon transfer reactions on targets that possess one-hole or few-hole configurations with respect to closed shells. Caution is needed in that such transfer reactions only probe a component of the many-nucleon structure of the intruder state. We note that this language is with respect to shell structure defined in a spherical mean field, but intruder structures are deformed. If a deformed structure is expanded in a spherical basis, the expansion is over a range of particle–hole configurations. The extent of this range is unknown.

The current extent of shape coexistence manifested in nuclei at low energy and low spin is presented in figure 4.44. The mass regions labelled A, ⋯ L cover the following nuclei presented herein:

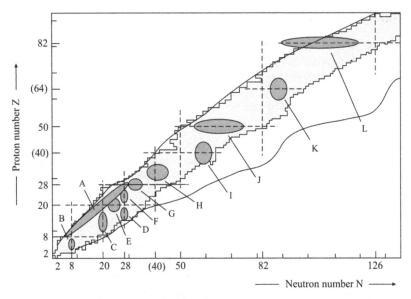

Figure 4.44. Global view of the systematic occurrences of shape coexistence in nuclei at low excitation energy and low spin. The labelled mass regions, A, ⋯ L are summarized in the text. Reprinted figure with permission from [37]. Copyright 2011 by the American Physical Society.

- A: $N \sim Z$, ^{16}O, ^{40}Ca (figures 3.1–3.4);
- B: $N \sim 8$, ^{11}Be (figure 4.43);
- C: $N \sim 20$, ^{32}Mg (Figures 3.17 and 4.27); Na isotopes (figure 3.16);
- D: $N \sim 28$ (figure 4.29);
- E: $Z \sim 20$, ^{42}Ca (figures 3.20–3.21 and 4.23); Sc isotopes (figure 4.22);
- F: $N \sim 28$ (figure 4.29);
- G: $Z \sim 28$, Cu isotopes (figure 4.30);
- H: $Z \sim 36$, Kr isotopes (figures 4.39–4.40);
- I: $N \sim 60$, ^{98}Zr (figures 4.1–4.2 and 4.4);
- J: $Z \sim 50$, Sn isotopes (figures 3.13, 3.23–3.26);
    ^{115,117}In (figures 3.5–3.7, 4.11–4.12 and 4.21);
    ^{110,112,114}Cd (figures 3.22, 3.25–3.27 and 4.37);
    Sb isotopes (figures 4.11–4.12, 4.21);
- K: $N \sim 90$, ^{152}Sm (figures 4.3–4.5, 4.8–4.10 and 4.38);
- L: $Z \sim 82$, Tl isotopes (figures 3.8–3.9, 3.15, 4.14–4.15 and 4.20);
    Hg isotopes (figures 3.10–3.12 and 3.18–3.19);
    Pb (figures 3.14 and 4.41–4.42);
    Bi (figures 3.15, 4.16 and 4.20);
    Au (figures 4.17–4.19);
    Pt (figures 4.32–4.33 and 4.36);
    Po (figure 4.35).

4.3 Is there a unified view of shape coexistence in nuclei?

The emerging view of shape coexistence in nuclei demands unification. Figure 4.45 depicts an attempt at such unification in terms of numbers of active protons and neutrons in relation to shell structure in nuclei. Caution is needed in adopting such a view because, where deformation emerges in nuclear structure, such shell structure is meaningless. Thus, the view is one of competition between competing schemes of organization of the nuclear many-body system. Further, the view presented in figure 4.45 is simplified to involvement of a few adjacent shells, when many shells are likely involved. To adopt a view that uses only one or a few shells in the interpretation of the emergence of deformation and multiple shapes in nuclei would be inadvisable. The evidence from electric quadrupole moments and transition strengths is that many-shells are active in generating the correlations behind the collectivity. This is manifested in the so-called 'effective charge' problem where, in shell model calculations it is necessary to assign charges, typically $+0.5e$ to neutrons and $+1.5e$ to protons. In the spirit of 'effective models', this is standard procedure. With respect to the emergence of collectivity in nuclei, the procedure buries any hope of understanding. This topic is explored at a rudimentary level in a recent review [36].

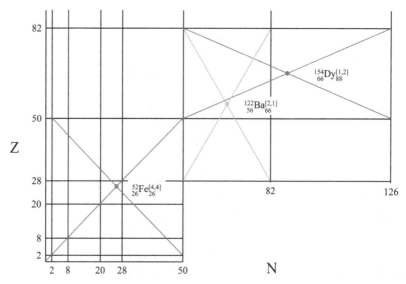

Figure 4.45. Illustration of the concept of 'multishells'. Removal of a closed-shell 'line' between two open-shell regions creates an open multi-shell region. For $Z = 82$ this provides an explanation of the extreme deformation associated with the coexisting states observed in the Hg and Pb isotopes. This perspective may also provide an explanation of the mass regions where superdeformation is observed. The superscripts, e.g., [2, 1] attached to $^{122}_{56}\mathrm{Ba}^{[2,1]}_{66}$, indicate the number of proton and neutron 'regular' shells forming the multishells as shown by the diagonal lines passing through the location of the isotope. The region in the lower left-hand corner contains mainly $N = Z$ lines cases; i.e., 'symmetric' cases; the region in the upper right-hand corner contains only asymmetric cases.

4.4 Exercises

4-1. With reference to figures 4.8 and 4.9, using data in ENSDF for ^{154}Gd, generate a similar band mixing analysis. Hint, make the heads of the two unmixed bands degenerate in energy, i.e., at spin zero. Two-band mixing analysis formalism, suited to this exercise, is presented in appendix B.

4-2. With reference to figure 4.33, using data in ENSDF, as far as possible, classify excited states in
   (a) $^{162-198}$Os;
   (b) $^{158-190}$W.

4-3. In figure 4.38(b), it is noted in the figure caption that there is a 631 keV transition observed in the conversion electron spectrum which has no counterpart in the gamma-ray spectrum. Consider possible explanations for this observation. Hint: it is significant that the 631 keV transition is supposed to connect the pair of $K = 2\ 3^+$ states at 1043 and 1674 keV.

A View of Nuclear Data

Nuclear shape coexistence
8. Shape coexistence and E0 transitions

E0 transitions are a key spectroscopic fingerprint of shape coexistence, where the coexisting structures undergo quantum mechanical configuration mixing.

E0 transitions are seen in association with shape coexistence established vie E2 properties.

E0 transitions can provide first indications of the presence of shape coexistence when mixing obscures E2 patterns

Tutorial 4.1 Shape coexistence and E0 transitions. The video can be downloaded from https://doi.org/10.1088/978-0-7503-5643-5.

A View of Nuclear Data

Mixing of simple model degrees of freedom
16. Mixing of two different deformed structures: ^{152}Sm

Two-state mixing of simple model structures is illustrated

The analysis achieves quantitative agreement with E2 data at the "1%" level

The analysis reveals that E2 strength, which might be interpreted as due to "intrinsic" (vibrational) dynamics, may be due entirely to simple mixing effects

Tutorial 4.2 Mixing of two different deformed structures: ^{152}Sm. The video can be downloaded from https://doi.org/10.1088/978-0-7503-5643-5.

References

[1] Kibédi T, Garnsworthy A B and Wood J L 2022 Electric monopole transitions in nuclei *Prog. Part. Nucl. Phys.* **123** 103930

[2] Cheifetz E, Jared R C, Thompson S G and Wilhelmy J B 1970 Experimental information concerning deformation of neutron-rich nuclei in the $A \sim 100$ region *Phys. Rev. Lett.* **25** 38

[3] Wang M, Huang W J, Kondev F G, Audi G and Naimi S 2021 The AME2020 atomic mass evaluation (II). tables, graphs and references *Chin. Phys. C* **45** 030003

[4] Angeli I and Marinova K P 2013 table of experimental nuclear ground state charge radii: an update *At. Data Nucl. Data tables* **99** 69

[5] Thibault C *et al* 1981 Hyperfine structure and isotope shift of the D_2 line of $^{76-98}$Rb and some of their isomers *Phys. Rev. C* **23** 2720

[6] Angeli I 2004 A consistent set of nuclear rms charge radii: properties of the radius surface $R(N, Z)$ *At. Data Nucl. Data tables* **87** 185–206

[7] Kulp W D *et al* 2007 Shape coexistence and mixing in ^{152}Sm (arXiv:0706.4129)

[8] Gaigalas A K, Shroy R E, Schatz G and Fossan D B 1975 Deformed $9/2^+$ states and $\Delta J = 1$ rotational bands in ^{113,115,117,119}Sb nuclei *Phys. Rev. Lett.* **35** 555

[9] Fossan D B, Gai M, Gaigalas A K, Gordon D M, Shroy R E, Heyde K, Waroquier M, Vincx H and Van Isacker P 1977 Deformed $9/2^+$ proton-hole states in odd-A I nuclei *Phys. Rev. C* **15** 1732

[10] Gordon D M, Gai M, Gaigalas A K, Shroy R E and Fossan D B 1977 Collective properties of $A = 117 - 127$ odd-A I nuclei *Phys. Lett. B* **67** 161–4

[11] Garg U, Sjoreen T P and Fossan D B 1978 Deformed $9/2^+$ proton-hole states on odd-A $^{119-125}$Cs *Phys. Rev. Lett.* **40** 831

[12] Garg U, Sjoreen T P and Fossan D B 1979 Collective properties of the odd-mass Cs nuclei. I. ^{127,129,131,133}Cs *Phys. Rev. C* **19** 207

[13] Garg U, Sjoreen T P and Fossan D B 1979 Collective properties of the odd-mass Cs nuclei. II. ^{119,121,123,125}Cs *Phys. Rev.* C **19** 217

[14] Shroy R E, Gaigalas A K, Schatz G and Fossan D B 1979 High-spin states in odd-mass $^{113-119}$Sb: $\Delta J = 1$ bands on $9/2^+$ proton-hole states *Phys. Rev.* C **19** 1324

[15] Shroy R E, Gordon D M, Gai M, Fossan D B and Gaigalas A K 1982 Collective properties of the odd-mass I nuclei: ^{123,125,127}I *Phys. Rev.* C **26** 1089

[16] Gai M, Gordon D M, Shroy R E, Fossan D B and Gaigalas A K 1982 Collective properties of the odd-mass I nuclei: ^{117,119,121}I *Phys. Rev.* C **26** 1101

[17] Fossan D B 1977 Rhodos conf. May 1977 *Technical report* LBL-2912

[18] Badran H *et al* 2017 Decay spectroscopy of ^{179}Pb and evidence for a $9/2^-$ intruder state in ^{179}Tl *Phys. Rev.* C **96** 064314

[19] Doherty D T *et al* 2021 Solving the puzzles of the decay of the heaviest known proton-emitting nucleus ^{185}Bi *Phys. Rev. Lett.* **127** 202501

[20] Sedlák M *et al* 2020 Nuclear structure of ^{181}Au studied via β+ /EC decay of ^{181}Hg at ISOLDE *Eur. Phys. J.* A**56** 161

[21] Jenkins D G and Wood J L 2021 *Nuclear Data: A Primer* (Bristol: IOP Publishing) https://iopscience.iop.org/book/mono/978-0-7503-2674-2

[22] Roussière B *et al* 2000 *High-resolution measurements of low-energy conversion electrons Hyperfine Interact* **129** 119

[23] Venhart M *et al* 2020 Population of a low-spin positive-parity band from high-spin intruder states in ^{177}Au: the two-state mixing effect *Phys. Lett.* B **806** 135488

[24] Joshi P, Kumar A, Mukherjee G, Singh R P, Muralithar S, Garg U, Bhowmik R K and Govil I M 2002 Configuration dependence of deformation in ^{183}Au *Phys. Rev.* C **66** 044306

[25] Joshi P, Kumar A, Govil I M, Singh R P, Mukherjee G, Muralithar S, Bhowmik R K and Garg U 2004 Deformation effects in ^{185}Au *Phys. Rev.* C **69** 044304

[26] Venhart M *et al* 2017 New systematic features in the neutron-deficient Au isotopes *J. Phys.* G **44** 074003

[27] Venhart M *et al* 2011 Shape coexistence in odd-mass Au isotopes: determination of the excitation energy of the lowest intruder state in ^{179}Au *Phys. Lett.* B **695** 82–7

[28] Venhart M *et al* 2017 De-excitation of the strongly coupled band in ^{177}Au and implications for core intruder configurations in the light Hg isotopes *Phys. Rev.* C **95** 061302

[29] Wood J L, Heyde K, Nazarewicz W, Huyse M and Van Duppen P 1992 Coexistence in even-mass nuclei *Phys. Repts.* **215** 101

[30] Bjerregaard J H, Hansen O, Nathan O, Chapman R, Hinds S and Middleton R 1967 The (t,p) reaction with the even isotopes of Ca *Nucl. Phys.* A **103** 33–70

[31] Petersen J F and Parkinson W C 1974 The ^{40}Ar(τ,n)^{42}Ca reaction *Phys. Lett.* B **49** 425–7

[32] Fortune H T, Betts R R and Bishop J N 1978 Location of 0^+ 4p-2h and 6p-4h configurations in ^{42}Ca *Nucl. Phys.* A **294** 208–12

[33] Peng J C, Stein N, Sunier J W, Drake D M, Moses J D, Cizewski J A and Tesmer J R 1979 Study of the reactions $^{46, 48}$Ti(^{14}C,^{16}O)$^{44, 46}$Ca and $^{50, 52}$Cr(^{14}C,^{16}O)$^{48, 50}$Ti at 51 MeV *Phys. Rev. Lett.* **43** 675

[34] Neyens G 2011 Multiparticle-multihole states in ^{31}Mg and ^{33}Mg: a critical evaluation *Phys. Rev.* C **84** 064310

[35] Neyens G 2016 Shape coexistence in the $N = 19$ and $N = 21$ isotones *J. Phys.* G **43** 024007

[36] Stuchbery A E and Wood J L 2022 To shell model, or not to shell model, that is the question *Physics* **4** 697–773

[37] Heyde K and Wood J L 2011 Shape coexistence in atomic nuclei *Rev. Mod. Phys.* **83** 1467

[38] Chiara C J *et al* 2015 Identification of deformed intruder states in semi-magic ^{70}Ni *Phys. Rev. C* **91** 044309

[39] Nowak K *et al* 2016 Spectroscopy of ^{46}Ar by the (t,p) two-neutron transfer reaction *Phys. Rev. C* **93** 044335

[40] Hinds S and Middleton R 1967 A study of the (t,p) reactions leading to ^{48}Ti and ^{50}Ti *Nucl. Phys. A* **92** 422–32

[41] Chapman R, Hinds S and MacGregor A E 1968 A study of ^{52}Cr, ^{54}Cr and ^{56}Cr by the (t,p) reaction *Nucl. Phys. A* **119** 305–24

[42] Casten R F, Flynn E R, Hansen O and Mulligan T J 1971 Strong $L = 0$ (t,p) transitions in the even isotopes of Ti, Cr and Fe *Phys. Rev. C* **4** 130

[43] Whitten C A 1967 Nuclear spectroscopy in Cr^{52}, Cr^{53}, and Cr^{54} by the (p,d), (p,t) and (p,p') reactions *Phys. Rev.* **156** 1228

[44] Suehiro T, Kokame J, Ishizaki Y, Ogata H, Sugiyama Y, Saji Y, Nonaka I and Itonaga K 1974 Core-excited states of ^{54}Fe from the ^{56}Fe(p,t)^{54}Fe reaction at 52 MeV *Nucl. Phys. A* **220** 461–76

[45] Nann H and Benenson W 1974 Levels of ^{56}Ni *Phys. Rev. C* **10** 1880

[46] Grévy S *et al* 2005 Observation of the 0_2^+ state in ^{44}S *Eur. Phys. J. A* **25** 111–3

[47] Force C *et al* 2010 Prolate-spherical shape coexistence at N = 28 in ^{44}S *Phys. Rev. Lett.* **105** 102501

[48] Santiago-Gonzalez D *et al* 2011 Triple configuration coexistence in ^{44}S *Phys. Rev. C* **83** 061305

[49] Hagberg E *et al* 1978 Staggering of the moments of inertia of very neutron-deficient platinum isotopes *Phys. Lett. B* **78** 44–7

[50] Heery J *et al* 2021 Lifetime measurements of yrast states in ^{178}Pt using the charge plunger method with a recoil separator *Eur. Phys. J. A* **57** 132

[51] Müller-Gatermann C *et al* 2018 Low-lying electromagnetic transition strengths in ^{180}Pt *Phys. Rev. C* **97** 024336

[52] Chen Q M *et al* 2016 Lifetime measurements in ^{180}Pt *Phys. Rev. C* **93** 044310

[53] Häfner G *et al* 2021 Lifetime measurements in ^{182}Pt using γ-γ fast-timing *Eur. Phys. J. A* **57** 174

[54] Walpe J C, Garg U, Naguleswaran S, Wei J, Reviol W, Ahmad I, Carpenter M P and Khoo T L 2012 Lifetime measurements in $^{182, 186}$Pt *Phys. Rev. C* **85** 057302

[55] Sels S *et al* 2019 Shape staggering of midshell mercury isotopes from in-source laser spectroscopy compared with density-functional-theory and Monte Carlo shell-model calculations *Phys. Rev. C* **99** 044306

[56] Day Goodacre T *et al* 2021 Laser spectroscopy of neutron-rich $^{207, 208}$Hg isotopes: illuminating the kink and odd-even staggering in charge radii across the N = 126 shell closure *Phys. Rev. Lett.* **126** 032502

[57] Day Goodacre T *et al* 2021 Charge radii, moments and masses of mercury isotopes across the N = 126 shell closure *Phys. Rev. C* **104** 054322

[58] Barzakh A *et al* 2012 Hyperfine structure anomaly and magnetic moments of neutron deficient Tl isomers with I = 9/2 *Phys. Rev. C* **86** 014311

[59] Barzakh A E *et al* 2013 Changes in the mean-square charge radii and magnetic moments of neutron-deficient Tl isotopes *Phys. Rev. C* **88** 024315

[60] Barzakh A E *et al* 2017 Changes in mean-squared charge radii and magnetic moments of $^{178-184}$Tl measured by in-source laser spectroscopy *Phys. Rev.* C **95** 014324

[61] Barzakh A E *et al* 2017 Onset of deformation in neutron-deficient Bi isotopes studied by laser spectroscopy *Phys. Rev.* C **95** 044324

[62] Barzakh A E *et al* 2018 Shell effect in the mean square charge radii and magnetic moments of bismuth isotopes near $N = 126$ *Phys. Rev.* C **97** 014322

[63] Barzakh A E *et al* 2019 Inverse odd-even staggering in nuclear charge radii and possible octupole collectivity in $^{217,\,218,\,219}$At revealed by in-source last spectroscopy *Phys. Rev.* C **99** 054317

[64] Barzakh A *et al* 2021 Large shape staggering in neutron-deficient Bi isotopes *Phys. Rev. Lett.* **127** 192501

[65] Barzakh A E *et al* 2016 Laser spectroscopy studies of intruder states in $^{193,\,195,\,197}$Bi *Phys. Rev.* C **94** 024334

[66] Cocolios T E *et al* 2011 Early onset of ground state deformation in neutron deficient polonium isotopes *Phys. Rev. Lett.* **106** 052503

[67] Seliverstov M D *et al* 2013 Charge radii of odd-A $^{191-211}$Po isotopes *Phys. Lett.* B **719** 362–6

[68] Cubiss J G *et al* 2018 Charge radii and electromagnetic moments of $^{195-211}$At *Phys. Rev.* C **97** 054327

[69] Kesteloot N *et al* 2015 Deformation and mixing of coexisting shapes in neutron-deficient polonium isotopes *Phys. Rev.* C **92** 054301

[70] Xu Y, Krane K S, Gummin M A, Jarrio M, Wood J L, Zganjar E F and Carter H K 1992 Shape coexistence and electric monopole transitions in ^{184}Pt *Phys. Rev. Lett.* **68** 3853

[71] Zhang W Q *et al* 2022 First observation of a shape isomer and a low-lying strongly-coupled prolate band in neutron-deficient semi-magic ^{187}Pb *Phys. Lett.* B **829** 137129

IOP Publishing

Nuclear Data
A collective motion view
David Jenkins and John L Wood

Chapter 5

Are there vibrations in nuclei?

Vibrations at low energy are considered from an historical and a global perspective. The most favored candidates for quadrupole vibrations, ^{110}Cd and its heavier neighbours, fail to match model patterns. The spectroscopic challenges of arriving at this view are emphasized. One-phonon octupole vibrations in spherical nuclei are sketched.

Concepts: harmonic vibrator model, quadrupole collectivity, octupole collectivity, spherical nuclei, liquid-drop models.

Learning outcomes: The key data view from this chapter is the illustration of how detailed spectroscopic data has cast doubt on the existence of low-energy quadrupole vibrations in nuclei. Notably, it is shown that studies of electric quadrupole matrix elements extending to the three-phonon level are essential—energy patterns alone have misled the nuclear structure community.

The issue of whether there are vibrations in nuclei divides into the occurrence of vibrations at low energy (0.5–10 MeV) and at high energy (10–40 MeV). High-energy vibrations in nuclei are unequivocal, they are the so-called giant resonance modes. They are resonances, i.e. they are not energy eigenstates, because they occur above the threshold for nucleon emission and lack centrifugal or Coulomb barriers for at least one decay channel. Low-energy vibrations have had a mixed reception as a recognized collective mode in nuclei. We present a description of the giant resonance modes later in the Series. Immediately one must subdivide the topic of low-energy vibrational modes in nuclei into the study of spherical nuclei and the study of deformed nuclei. As with rotations, it is essential to adopt guidance from simple models. Such models differ substantially between those for spherical nuclei and those for deformed nuclei. Herein, we limit ourselves to the simplest possible spherical models. We address vibrations in deformed nuclei in the following chapter.

5.1 Historical view of low-energy vibrations in nuclei

The idea that nuclei support low-energy vibrational modes originated with the seminal work of Aage Bohr in 1952 [1]. Bohr proposed a liquid drop model of the nucleus, which could be spherical and exhibit quantized vibrations, or it could deform and exhibit quantized rotations and vibrations.

Experimental data in support of low-energy quadrupole vibrations in nuclei was published already in 1955 by Gertrude Scharff-Goldhaber and Joseph Weneser [2]. This view was broadened in 1960 by Raymond Sheline [3] with an assessment of evidence for low-energy vibrations in deformed nuclei, albeit with a severe lack of useful data for the task.

The simplest model of vibrations is depicted in figure 5.1. This treats the nucleus as a uniform liquid drop with a sharp surface. This droplet of fluid can deform, maintaining constant density. The excitations have a simple 'phonon' structure and are equidistantly spaced. More subtly, the transition strength from the two-phonon state to the one-phonon state is twice the strength from the one-phonon state to the ground state. The transition strength rule follows from the elementary quantum mechanics of the one-dimensional simple harmonic oscillator, expressed in algebraic terms, viz.

$$a^{\dagger}|0\rangle = |1\rangle \tag{5.1}$$

$$a^{\dagger}|n\rangle = \sqrt{n+1}\,|n+1\rangle \tag{5.2}$$

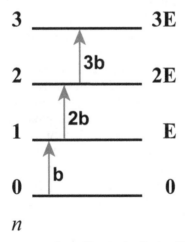

Figure 5.1. Schematic view of a quantum mechanical harmonic vibrator. The excitations are characterized by equidistant spacing of energy levels: this is well known from the elementary nature of the quantum mechanical one-dimensional harmonic oscillator. The transition strengths are characterized by a monotonic increase between these levels: this is less-well known, and the reason is presented in the algebra shown in the text. The only labelling quantum number at this level of detail is n, the number of oscillator 'phonons'. The transition strengths are expressed as matrix elements, which are proportional to the operator connecting the states. Note, the transitions are shown as excitations, but the matrix elements are the same for de-excitation.

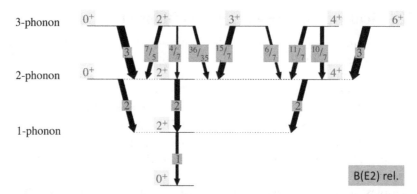

Figure 5.2. Schematic view of the harmonic quadrupole vibrator, i.e. for oscillator phonons carrying a spin of 2. Such phonons are bosons and the allowed total spins for multi-phonon coupling are determined using the m-scheme (see exercises 5-1 and 5-2). The transition strengths follow the naïve view presented in figure 5.1 up to the two-phonon level, but at the three-phonon level the $n = 3 \rightarrow n = 2$ transition strengths reflect partitioning due to the spin structure of the phonons, which is beyond the present level of discussion (note that the partitions sum to three units of oscillator strength for each member of the three-phonon quintuplet).

$$T(n \rightarrow n + 1) = b\langle n + 1|a^\dagger|n \rangle^2 = b(n + 1) \tag{5.3}$$

For spherical nuclei, the leading order vibrational mode is quadrupolar and this is depicted in figure 5.2. This mode is treated as a harmonic oscillator with quanta carrying a spin of two. Thus, the one-phonon excitation has spin 2, and two-phonons of excitation result in three excited states with spins[1] 0, 2 and 4. The two-phonon triplet forms a degenerate multiplet at twice the energy of the one-phonon excitation. Figure 5.2 provides a useful view of the key energy signatures and $B(E2)$ signatures for assessing low-energy quadrupole vibrational modes in nuclei. Specifically, candidate nuclei should exhibit $E(4_1)/E(2_1) \sim 2$ and $B_{42}/B_{20} \sim 2$. There appear to be some candidates. Indeed, based on these two criteria, together with the phonon selection rule which dictates that $B_{2'0}/B_{2'2} \sim 0$, an extensive body of literature developed such that it was widely accepted: 'nuclei near closed shells are spherical and exhibit low-energy quadrupole vibrations'. However, little attention was paid to the requirement that $B_{2'2} \sim 2B_{20}$ and $B_{0'2} \sim 2B_{20}$. This lack of attention was rationalized on the basis that lifetime measurements in the ps–ns range had to be made. Such measurements are, in general, not easy. Further, $2' \rightarrow 2$ transitions necessitated the determination of the $E2/M1$ mixing ratio for the transition. This required $\gamma\gamma$ angular correlation measurements, i.e. coincidence measurements as a function of the angle between two detectors. Again, such measurements are not easy. Yet further was the impediment that excited 0^+ states are difficult to populate, especially using in-beam methods where lifetime information is accessible by Doppler shift methods.

---

[1] See exercise 5-1 for the use of the m-scheme (for bosons) applied to the deduction of these spins for the two-phonon multiplet. Also see exercise 5-2 for the deduction of spins 0, 2, 3, 4, 6 for the three-phonon multiplet.

The above view prevailed for fifty years. It is fair to say that there was universal satisfaction with this view being correct. In the early 2000s a program at University of Kentucky was underway to characterize the best candidate nuclei for harmonic quadrupole vibrations. There were 'textbook' cases: ^{110}Cd was a favoured example [4]. The studies at University of Kentucky focussed on gamma-ray spectroscopy following inelastic scattering of neutrons from ^{110,112,114,116}Cd. These studies provided a comprehensive view of the low-energy excitations and, most importantly, lifetime information in the 30–700 fs range from Doppler energy shifts of gamma rays. The revelation came with respect to the candidate states for three-phonon excitations: the expected 3-phonon → 2-phonon $B(E2)$ strength, which should sum to $3B_{20}$, was not realized in these nuclei. This turned an optimistic exploration of low-energy harmonic quadrupole vibrations into a realization that the concept was questionable, i.e. low-energy multi-phonon excitations may not exist in nuclei for quadrupole excitations. Key steps in this saga are described in [5, 6]. We look at the pros and cons for low-energy quadrupole vibrations in nuclei in the next section.

5.2 Assessment of low-energy quadrupole vibrations in nuclei

The isotopes ^{110,112,114,116}Cd are generally considered the textbook example of low-energy quadrupole vibrations in nuclei. The details of $E2$ transition strengths between the low-energy excited states in ^{110,112,114}Cd are presented, in comparison to the pattern in figure 5.2, in figures 5.3(a)–(d). The most problematic feature for a vibrational interpretation is the extremely weak $B(E2)$ value from the candidate 2-phonon 0^+ state to the 1-phonon (2^+) state in ^{112,114}Cd. Further, the $B(E2)$ values from the candidate 3-phonon 2^+ state to the candidate 2-phonon 2^+ and 4^+ states are very weak in all three isotopes. Indeed, one notes that the candidate 3-phonon 2^+ state exhibits a strong $B(E2)$ value to the candidate 2-phonon 0^+ state in all three isotopes: this identifies the two states, unequivocally in ^{112,114}Cd and by implication in ^{110}Cd, as forming the beginning of a band structure. This band is manifestly isolated from all the other low-energy states in all these isotopes.

Figures 5.4(a)–(d) shows a rearrangement of the states depicted in figures 5.3(a)–(d), together with the low-energy shape coexisting, intruder states, cf 3.22, to reflect evident multiple band structures in ^{110,112,114}Cd. Most notably, the $B(E2)$ values support the beginning of a $K = 2$ band, constituted from the candidate 2-phonon 2^+ state and the candidate 3-phonon 3^+ and 4^+ states. The ground state is the head of a $K = 0$ band and the above-noted newly emerging band can be interpreted as having $K = 0$ and is very similar to the ground-state band. The shape-coexisting states form a $K = 0$ band as already identified in 3.22. Yet further, as shown in 4.37, ^{114}Cd exhibits the characteristics of $K = 2$ coexisting bands, and the present identification sharpens the view of the lowest $K = 2$ band. We particularly note that $E0$ transitions play a key role in this structural identification: $E0$ transitions cannot change K and a $\Delta K = 0$ selection rule is evident for the patterns of $E0$ transitions in 4.37. As a capstone perspective on this rearrangement to the low-energy structure of the even Cd isotopes, figures 5.5(a) and (b) shows the population of low-lying states in ^{108,110}Cd by the ^{107,109}Ag(^{3}He,d) reaction. Evidently, the 0^+ state in ^{110}Cd that

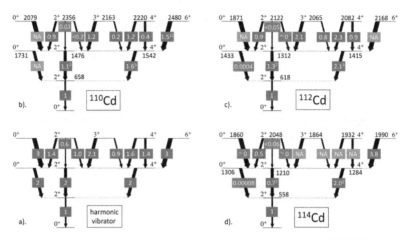

Figure 5.3. States that are candidates for quadrupole vibrational behaviour in ^{110,112,114}Cd. The data are arranged to match the pattern shown in figure 5.2, which is repeated in frame (a). Frames (b), (c), and (d) summarize currently available data for ^{110}Cd, ^{112}Cd and ^{114}Cd, respectively. Energies are given in keV and $E2$ transition strengths are given relative to $B(E2; 2_1^+ \rightarrow 0_1^+) = B_{20} = 1.00$; the B_{20} values in W.u. (from ENSDF) are 27.08 (^{110}Cd), 30.32 (^{112}Cd) and 31.2 (^{114}Cd). Transitions for which data are not available are labelled 'NA'. Severe departures from expectations for harmonic quadrupole vibrational behaviour are highlighted in red; note that these exhibit a systematic pattern.

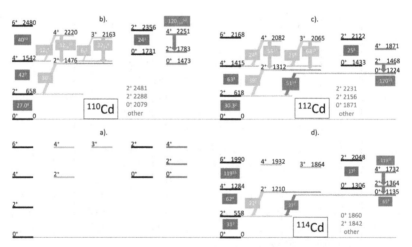

Figure 5.4. Low-energy excited states in ^{110,112,114}Cd arranged into band patterns and contrasted with a harmonic quadrupole vibrator view. The data are arranged to match the pattern shown in figure 5.2, which is rearranged in frame (a). Frames (b), (c), and (d) summarize currently available data for ^{110}Cd, ^{112}Cd and ^{114}Cd, respectively. Energies are given in keV. The $E2$ transition strengths are given in W.u. and are shown only for strong transitions, where known. Colour coding is simply to assist in recognizing the groupings of the $E2$ transitions. Other, lowest positive-parity excited states that are omitted are listed by energy. The bands conform to patterns for $K = 0, 2$ and are discussed further in the text. The data are taken from ENSDF and the $B(E2)$ values are rounded to essential significant figures.

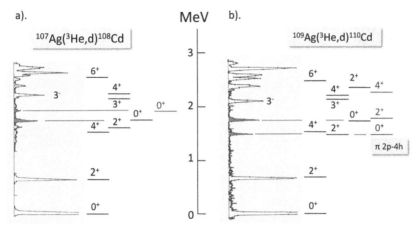

Figure 5.5. States in ^{108,110}Cd populated in the ^{107,109}Ag(^{3}He,d)^{108,110}Cd one-proton transfer reaction. The spectra of proton ejectiles are matched to the low-energy excited states in the final nuclei, arranged into K bands as in figure 5.4, with excited 0^+ states highlighted in red. Note that the one-proton transfer strength is fragmented between the intruder $K = 0$ band head, labelled π 2p–4h, and the head of the $K = 0$ band proposed in figure 5.4, with somewhat more strength to the proposed (non-intruder) band. Population of first excited 3^- states is indicated. Other features in the proton ejectile spectra involve excited states which are not relevant to the present discussion. Further details are discussed in the text. The data are taken from [7] and from ENSDF.

has been considered as a 2-phonon state has long been known to have significant proton excitation character, possibly a redistribution of $J = 0$ proton pairs within the $1g_{9/2}$–$2p_{1/2}$ subshells.

A further feature of these cadmium isotopes is that the 2^+_1 states are observed to have non-zero quadrupole moments, viz. -0.40^3 b (^{110}Cd); -0.38^3 b (^{112}Cd); -0.35^5 b (^{114}Cd), cf ENSDF. These values imply that the cadmium isotopes are deformed. Historically, these values were addressed as evidence of anharmonic vibrations. We do not consider such descriptions of the cadmium isotopes herein because of the complexity of such descriptions of nuclei. The present view is strictly for harmonic quadrupole vibrations of spherical nuclei, for which $Q = 0$ for all states.

The data presented here capture the essential steps in the historical realization [5, 6] that these Cd isotopes do not provide a textbook example of low-energy quadrupole vibrational behaviour. To the contrary, they exhibit quasi-rotational bands that can be labelled by K quantum numbers. Thus, the title of this chapter comes into focus. One way forward is suggested by the details presented in figure 5.1: one must thoroughly characterize the $E2$ transitions in nuclei with $R_4 \sim 2.0$ and $B_{42}/B_{20} \sim 2.0$. Figures 5.6(a) and (b) present a comprehensive view of R_4 and B_{42} versus B_{20} for all nuclei with $B_{20} < 100$ W.u.

The challenge to exploring excited states in nuclei such as $^{110-116}$Cd is the assignment of higher lying states into bands or multi-phonon multiplets. The issue is the difficulty in determining the intensities of low-energy transitions between high-lying states. This is due to the ever-present issue of the E^5_γ factor in observed $E2$

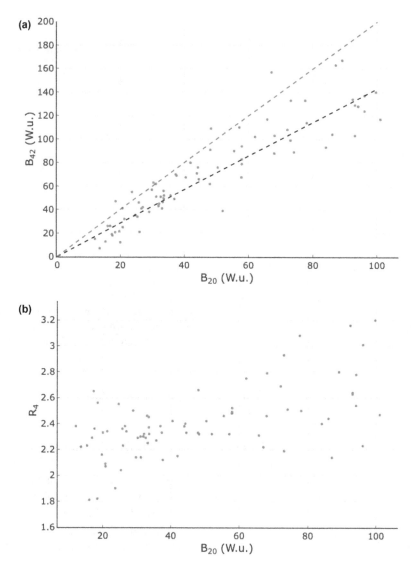

Figure 5.6. (a) Plot of B_{42} versus B_{20} for all nuclei with $B_{20} < 100$ W.u. The red dashed line indicates the expected trend for vibrational nuclei while the black dashed line indicates the expected trend for rotational nuclei. (b) Plot of R_4 versus B_{20} for all nuclei with $B_{20} < 100$ W.u. The limit of 100 W.u. on B_{20} is arbitrarily set as a lower limit for nuclei that might be considered to possess collective rotations; thus, the nuclei included in these plots should cover all cases where collective vibrations might be considered as manifested. Error bars are suppressed for clarity of presentation but the reader should bear in mind that many of transition strengths, in particular, have significant uncertainty. (Interactive version available for e-book can be downloaded from http://iopscience.iop.org/book/mono/978-0-7503-5643-5.)

transition decay rates. An example is shown in figure 5.7. The only solution is to take data using large arrays of detectors running for very long data acquisition times at very high counting rates. All three of these factors have limitations: cost of detectors, human cost in working hours, equipment sharing limitations, event acquisition-rate

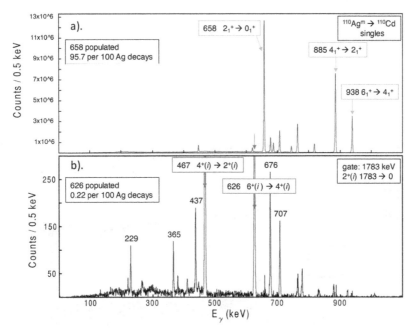

Figure 5.7. Illustration of the observation and quantification of very weak decay branches in the nucleus ^{110}Cd following the radioactive decay ^{110}Ag^m (β^-, $T_{1/2} = 249.83$ d, $J^\pi = 6^+$, $Q_\beta = 2891$ keV). These weak decay branches occur uniformly between excited states in all nuclei, where the E_γ^5 factor, inherent in $E2$ transition rates, dominates radiative decay. (a) The singles spectrum shows identification of some strong transitions, cf figure 5.4(b); the red arrow shows the barely visible 626 keV peak. (b) Spectrum of γ rays in coincidence with the 1783 keV γ ray de-exciting the 1783 keV, $J^\pi = 2^+$ member of the intruder band, $2^+(i)$ to the ground state. Notably, the spectrum shows a 467 keV γ ray which supports the intruder band structure shown in figure 5.4(b) (and see 3.22). The 1783 keV γ-ray transition occurs 1.02×10^{-4} per ^{110}Ag^m β decay. The 467 keV γ ray occurs 2.5×10^{-4} per ^{110}Ag^m β decay yet corresponds to a transition with a strength of 120_{110}^{50} W.u. The spectrum shows many other γ-ray lines corresponding to transitions which feed the 1783 keV state directly and indirectly. The figure is provided courtesy of J M Allmond and the data were acquired in a 200 day data collection with the CLARION array at Oak Ridge National Laboratory. The figure is similar to one in [8].

limitations of electronics. *This aspect of nuclear spectroscopic study has become one of the major limitations to advancing our understanding of nuclear structure.* We address the other limitations shortly.

It is evident that what is needed, to advance our understanding of nuclei such as the Cd isotopes and their collectivity, is a 'map' of the electric quadrupole excitation response such as can be achieved in multi-step Coulomb excitation. This is illustrated in figure 5.8. With respect to the 'E_γ^5 problem', there is a positive view. This factor applies to the spontaneous radiative decay of the state in question; it does not apply to Coulomb excitation. While one observes the result of Coulomb excitation via the relaxation of the nucleus back to the ground state by spontaneous radiative decay, these gamma-ray yields are deconvoluted into what happened in the multi-step excitation paths. In such paths, low-energy collective transitions have a

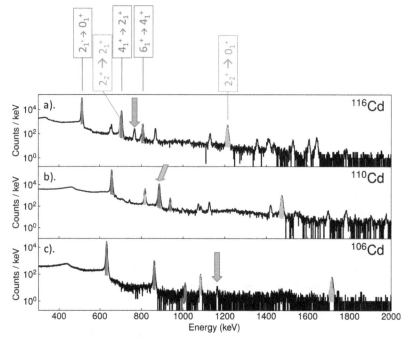

Figure 5.8. Coulomb excitation of ^{106,110,116}Cd beams on a ^{208}Pb target, observed via γ-ray spectroscopy. The spectra shown, (a) ^{116}Cd, (b) ^{110}Cd and (c) ^{106}Cd, were obtained under identical conditions, thus permitting a direct comparison of the multi-step electromagnetic responses of these nuclei. The transitions assigned to the observed strong γ rays are shown, cf figure 5.4(b) (and 3.22) for ^{110}Cd. The (locations of) $0_2^+ \rightarrow 2_1^+$ transitions are indicated with bold orange arrows. Note that the $2_2^+ \rightarrow 2_1^+$ and $0_2^+ \rightarrow 2_1^+$ transitions are an unresolved doublet in ^{110}Cd. Other γ-ray lines in the spectra are considered in exercise 5-7. The figure is provided courtesy of T J Gray and J M Allmond. The data were acquired with GRETINA-CHICO-2 at Argonne National Laboratory. The data for ^{106}Cd are presented in [9].

high probability, with a much less sensitive dependence on how far removed they are from the ground state (but ultimately only a finite number of excitation steps occur in an experiment).

A feature of multi-step Coulomb excitation studies, that can be confusing to the inexperienced eye, is the deduction of strengths ($E2$ matrix elements) for transitions that are not observed. The reason they are not observed is because observation is via relaxation after the Coulomb excitation process; this is via the process of spontaneous emission which is subject to the E_γ^5 factor. The reason they can be deduced is because they play a dominant role in the Coulomb excitation process. A dramatic manifestation of this is the population of negative-parity states: the excitation mode is via $E3$ matrix elements, the de-excitation mode is via $E1$ radiative decay. Excitation by $E1$ matrix elements has low probability because such matrix elements for nuclei are very small between low-energy states; de-excitation by $E3$ radiative decay has low-probability because coupling to the EM vacuum is weak due to the high spin change.

5.3 Low-energy octupole vibrations in spherical nuclei

Octupole vibrations in nuclei have been a topic of much less interest than quadrupole vibrations, because they are not dominant. There is a candidate one-phonon octupole state in most even–even nuclei if that nucleus has been studied in any detail. A map of excitation energies of 3_1^- states is shown in figure 5.9. The $B(E3)$ strength associated with the 3^- states in figure 5.9 is shown in figure 5.10. Such excitations are directly observed by inelastic scattering. Examples are shown for the double-closed shell nucleus, ^{208}Pb in figure 5.11 and for the single-closed shell nucleus, ^{118}Sn in figure 5.12. Note there is a major review of this topic [10].

The question of the existence of two-phonon octupole excitations remains elusive. Two technical factors are: they will have excitation energies where the level density is high and they will have positive parity; therefore, they will be difficult to characterize against a 'background' of many excited states, many poorly characterized, and they will decay by transitions with $E2$ and $M1$ multipolarity. There has been some interest in excitations described as one-phonon quadrupole coupled to one-phonon octupole, notably in the Cd isotopes. But such excitations are now in question, based on details given in section 5.2.

The occurrence of octupole collectivity as a function of N and Z does not match liquid drop descriptions. A liquid drop should exhibit the lowest energy and highest strength for a collective excitation far from closed shells where the nucleus is least 'rigid'. This indicates that octupole collectivity is controlled by properties of proton and neutron configurations. This point is visited in detail for deformed nuclei.

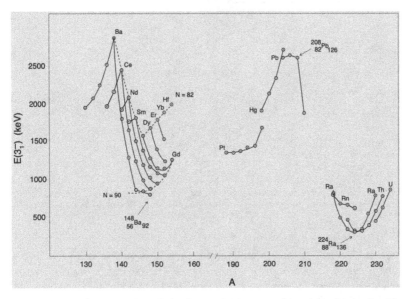

Figure 5.9. Systematics of the energies of the first excited states with spin-parity 3^- for $56 \leqslant Z \leqslant 64$, $78 \leqslant Z \leqslant 82$, and $88 \leqslant Z \leqslant 92$. Isotopes are joined by solid lines and isotones by dashed lines. The data are taken from [10]. The figure is reproduced from [11], copyright 2010 World Scientific Publishing Company.

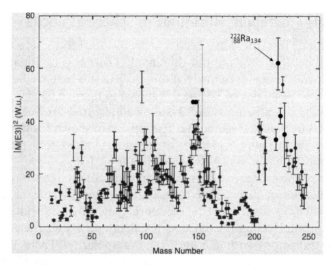

Figure 5.10. Systematics of electric octupole transition strength $|M(E3)|^2$ between the ground state and first excited 3^- state in doubly-even nuclei. The $|M(E3)|^2$ values are in Weisskopf units, W.u. The figure provides an update to data earlier presented in [10]. The figure is reproduced from [11], copyright 2010 World Scientific Publishing Company.

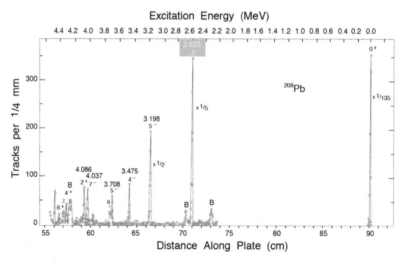

Figure 5.11. Inelastic deuteron scattering from ^{208}Pb. The data are for an incident energy of 13.1 MeV and a scattering angle of 150° with respect to the incident beam. Deuteron lines are labelled by the spins and parities of the corresponding levels. Lines marked with a B are 'background' events resulting from target contaminants. Note the scale reduction factors. The energies and spin-parities are from ENSDF. The figure is reproduced from [11], copyright 2010 World Scientific Publishing Company.

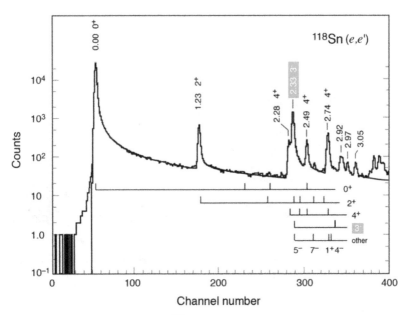

Figure 5.12. Inelastic electron scattering from ^{118}Sn. The data are for an incident energy of 252 MeV and a scattering angle of 68° with respect to the incident beam. The peaks corresponding to excited states reveal those states which are connected to the ground state by the largest electromagnetic matrix elements. The $I^\pi = 2^+$ state at 1.23 MeV and the $I^\pi = 3^-$ state at 2.33 MeV are deduced to be excited by strong electric quadrupole and electric octupole processes, respectively. The location of other excited states in ^{118}Sn is indicated; spins, parities, and excitation energies (in MeV) are taken from ENSDF. The figure is discussed further in the text. Reprinted figure with permission from [12], copyright 1991 by the American Physical Society.

5.4 Exercises

5-1. Deduce allowed spins for two spin-2 bosons using the m scheme.

5-2. Repeat exercise 5-1 for three spin-2 bosons.

5-3. Using the algebra presented in equations (5.1)–(5.3), show that a $n \to n - 2$ transition is impossible within the harmonic vibrator model (cf figure 5.1).

5-4. Deduce allowed spins for two spin-3 bosons using the m scheme.

5-5. With reference to figures 5.3 and 5.4, using data in ENSDF, as far as possible, classify excited states in
   (a) ^{118,120,122}Te;
   (b) ^{104,106,108}Pd.

5-6. Make use of data in ENSDF to identify candidate nuclei with $J = 0, 2, 4$ triplets which lack lifetime data.

5-7. With reference to figure 5.8, using data in ENSDF, identify the γ-ray lines in the spectra associated with the decays of:
   (a) the 3_1^- states;
   (b) the 0_3^+ states;
   (c) other states in ^{116}Cd.

A View of Nuclear Data

Low-energy nuclear vibrations
9. Spherical nuclei: where are the quadrupole vibrations?

Presentation of key data are made that question the concept of low-energy quadrupole vibrations in nuclei.
The critical issue is the existence of multi-phonon states.

Cadmium isotopes:
based on energy patterns (the historical view), multi-phonon excitations appeared to exist;
based on E2 transition strengths, multi-phonon excitations are unequivocally refuted.

Tellurium isotopes (for example):
current data are inadequate to answer the question;
but existing spectroscopic information raises interesting questions

Tutorial 5.1 Spherical nuclei: where are the quadrupole vibrations?. The video can be downloaded from https://doi.org/10.1088/978-0-7503-5643-5.

A View of Nuclear Data

Comprehensive (complete) spectroscopy
17. Studies of nuclei by (n,n'γ)

The unique capabilities of (n,n' γ) spectroscopy are illustrated, specifically:
a comprehensive mapping of excitations in nuclei (for spins <6)
location of γ-rays in a level scheme
determination of level lifetimes and spins
determination of transition multipolarities

A first view of deduced structure patterns by such spectroscopy is presented

NOTE: such measurements require gram quantities of highly enriched isotopes. Generally, this is done on a lease basis; but can still cost ~ 10-100 k$. Clearly, only stable isotopes can be studied.

Tutorial 5.2 Comprehensive spectroscopy: Studies of nuclei by (n,n'γ). The video can be downloaded from https://doi.org/10.1088/978-0-7503-5643-5.

References

[1] Bohr A 1952 The coupling of nuclear surface oscillations to the motion of individual nucleons *Mat. Fys. Medd. Dan. Vid. Selsk.* **26** (14)

[2] Scharff-Goldhaber G and Weneser J 1955 System of even-even nuclei *Phys. Rev.* **98** 212

[3] Sheline R K 1960 Vibrational states in deformed even-even nuclei *Rev. Mod. Phys.* **32** 1

[4] Rowe D J 2010 *Nuclear Collective Motions: Models and Theory* (Singapore: World Scientific)

[5] Garrett P E and Wood J L 2010 On the robustness of surface vibrational modes: case studies in the Cd region *J. Phys. G Nucl. Part. Phys.* **37** 064028

[6] Garrett P E, Wood J L and Yates S W 2018 Critical insights into nuclear collectivity from complementary nuclear spectroscopic methods *Phys. Scr.* **93** 063001

[7] Auble R L, Horen D J, Bertrand F E and Ball J B 1972 Proton excitations in $^{108, \, 110}$Cd from (^{3}He,d) reactions *Phys. Rev.* C **6** 2223

[8] Allmond J M 2016 Investigating shape evolution and the emergence of collectivity through the synergy of Coulomb excitation and β decay *EPJ Web of Conf.* **123** 02006

[9] Gray T J *et al* 2022 $E2$ rotational invariants of 0_1^+ and 2_1^+ states for ^{106}Cd: the emergence of collective rotation *Phys. Lett.* B **834** 137446

[10] Kibédi T and Spear R H 2002 Reduced electric-octupole transition probabilities, $B(E3; \, 0_1^+ \rightarrow 3_1^-)$–an update *At. Data Nucl. Data tables* **80** 35

[11] Rowe D J and Wood J L 2010 *Fundamentals of Nuclear Models: Foundational Models* (Singapore: World Scientific)

[12] Peterson R J, Kraushaar J J, Braunstein M R and Mitchell J H 1991 Inelastic electron scattering to collective states of ^{118}Sn *Phys. Rev.* C **44** 136

IOP Publishing

Nuclear Data
A collective motion view
David Jenkins and John L Wood

Chapter 6

Are there vibrations in deformed nuclei?

Low-energy vibrations in deformed nuclei are considered for quadrupole, octupole and hexadecapole degrees of freedom. Microscopic structure of octupole vibrations is introduced.

Concepts: deformed nuclei, quadrupole vibrations, octupole vibrations, deformed liquid drop models, hexadecapole collectivity.

Learning outcomes: The key data view from this chapter is the exploration of possible vibrational structures in deformed nuclei and the limitations of such a view, namely that such structures at best are confined to one-phonon excitations. The evidence for a hexadecapole degree of freedom is summarized, probably this is a permanent hexadecapole deformation.

Deformed nuclei exhibit a variety of excited states that have been considered as one-phonon excitations; and by implication searches for two-phonon excitations have been made. The topic has received much attention because it is possible to classify large numbers of excited states in strongly deformed even–even nuclei in terms of a limited number of rotational bands. Then one can look at the population of subsets of states by inelastic scattering of charged particles, Coulomb excitation, and one-nucleon transfer reactions. Also, with respect to excited 0^+ states, two-nucleon transfer reactions play an important role in probing the role of pairing in excited states. Further, by use of the Nilsson model, it is possible to suggest intrinsic structures for most of these bands in terms of 'broken pair' states involving Nilsson configurations.

A leading spectroscopic probe of vibrations in deformed nuclei has been inelastic scattering of deuterons. Indeed, this was the historic choice, manifested in the program of Bent Elbek at Niels Bohr Institute, Copenhagen[1], which laid the

---

[1] The results of this program, pertinent to the present discussion, were published in [1] (Sm isotopes), [2] (Gd isotopes), [3] (Dy isotopes), [4] (Er isotopes), [5] (Yb isotopes), and [6] (W isotopes). The Hf isotopes were studied but never published (see [7]).

doi:10.1088/978-0-7503-5643-5ch6

groundwork for much of our modern view of vibrations in deformed nuclei. An example of inelastic scattering of deuterons on a deformed nucleus, [168]Er is depicted in figure 6.1. A comprehensive view of the lowest excited states in [168]Er, organized into rotational bands, is presented in figure 6.2. The first feature to note is that only a few of the excited states in [168]Er are strongly populated by inelastic scattering. The second feature to note is that all the strongly populated excited states have $J^\pi = 2^+$, 3^-, 4^+ and the 6^+ member of the ground-state band. A simple and useful interpretation of this set of J^π values can be made: the dominant population of 2^+ states is a manifestation of quadrupole collectivity; the significant population of

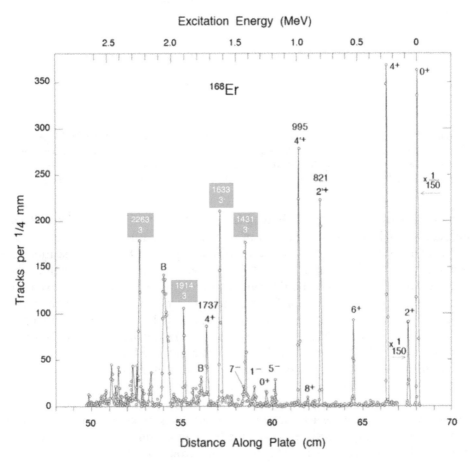

Figure 6.1. Population of excited states in [168]Er by inelastic scattering of deuterons. The data are for an incident energy of 12.098 MeV and a scattering angle of 125° with respect to the incident beam. Events are counts per 1/4 mm interval as a function of distance in cm along a photographic plate in the focal plane of a magnetic spectrometer, cf chapter 6, figure 6.13 in [8]. Deuteron lines are labeled by the spins and parities of the corresponding levels (cf figure 6.2). Lines marked with a B are background events resulting from target contaminants. Note the scale reduction factor for the ground state and first excited 2^+ state. The strongly populated states are indicated in figure 6.2 by red triangles. The spin-parities and energies of states are from ENSDF. The original data for this figure are derived from [4]. The figure is reproduced from [9], copyright 2010 World Scientific Publishing Company.

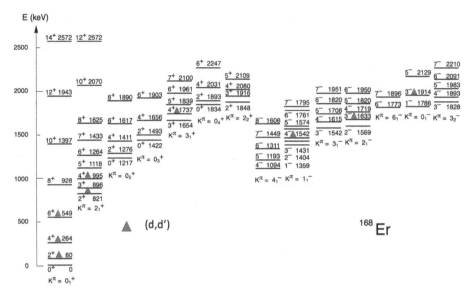

Figure 6.2. Excited states in the even–even nucleus, ^{168}Er organized into rotational bands. This illustrates the power of the rotational model to organize large numbers of excited states into a much smaller number of band structures. The states populated by the inelastic scattering of deuterons, (d,d'), cf figure 6.1, are indicated with red triangles. The figure is reproduced from [9], copyright 2010 World Scientific Publishing Company.

3^- states signals octupole collectivity; the population of 4^+ states and the 6^+ state reflects two- and three-step excitation of a quadrupole nature.

Figure 6.3 provides a systematic view of states populated by inelastic scattering of deuterons in ^{160,166,168}Dy and ^{160,166,168}Er. The ground state band in each of these nuclei is strongly populated and this can be inferred as due to multi-step excitation. Noting the counts-scale reduction factors, there is a weak but clear population of the 2^+ and 4^+ members of the so-called gamma band in each of these nuclei. Figure 6.4 provides a view of the collective states expected for the quadrupole normal modes of a spheroidal (football-shaped) liquid drop, as first proposed by Aage Bohr [10], and developed by Bohr and Mottelson [11]. We note that all identifications of spins and parities in figure 6.3 depend on detailed spectroscopy for each of these isotopes using other spectroscopic techniques; but some indications of spins for populated states were obtained in these inelastic scattering experiments by comparing intensities of peaks at different angles with respect to the beam direction (90° and 125° were used in the Elbek program). Also, inelastic scattering favours population of natural parity states, i.e., spin-parity 0^+, 1^-, 2^+, 3^-, 4^+, etc. The spectra shown here are for 125°; the larger the angle the more that the population of higher-spin states is favoured. Other members of the gamma bands, notably the 3^+ states, are not observably populated.

We follow below with the discussion of these and other data by focussing on various identifiable excitation modes that are candidates for vibrational degrees of freedom in deformed nuclei.

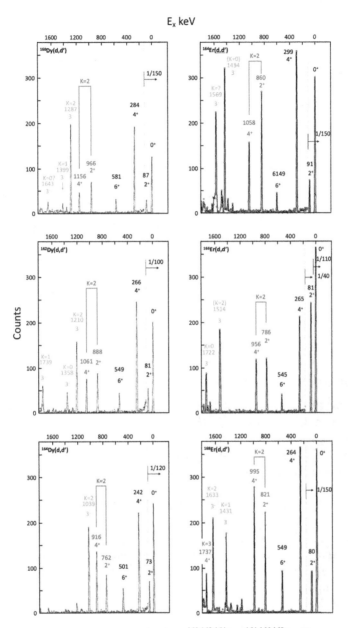

Figure 6.3. Deuteron spectra for inelastic scattering from ^{160,162,164}Dy, ^{164,166,168}Er. The spectra were taken at a bombarding energy of 12.1 MeV, at an angle of 125° with respect to the beam. Note the scale reduction factors. Excitation energies of the populated states are given in keV. The $K = 2$ gamma band 2$^+$ and 4$^+$ members are indicated in red. The 3$^-$ states populated are indicated in green, with K quantum number assignments (see text). Note the variability in the relative population of the 2$^+$ and 4$^+$ members of the $K = 2$ gamma band. This and other features are discussed in the text. Spectra reprinted from [3] and [4] with permission from Elsevier.

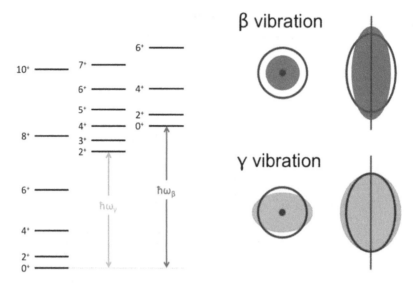

Figure 6.4. The normal modes of shape vibration in a spheroidal body and the bands associated with one-phonon excitations.

6.1 Gamma and beta vibrations

Gamma bands are observed in all deformed nuclei, without exception. Figure 6.5 presents a systematic view of the 2^+, $K = 2$ gamma band-head energies for all nuclei in the rare-earth region for which $E2$ strengths have been determined, bounded by $N > 90$ and $Z < 76$. These $E2$ strengths are shown in figure 6.6. The bounds are dictated by nuclei with exclusion of $N = 90$ isotones because they exhibit shape coexistence and exclusion of $Z = 76$ (osmium) isotopes because they exhibit characteristics of axially asymmetric rotors: thus, they are discussed in chapters 3 and 2, respectively. Figure 6.6 also shows $E2$ decay strengths for the 2^+ states in the lowest excited $K = 0$ bands in these nuclei.

The popular view of the $K = 2$, gamma band and the lowest $K = 0$ excited band (often called the beta band) is that they are normal-mode vibrations of a spheroidal liquid drop, cf figure 6.4. The data view presented here raises some serious questions about a liquid drop view:

1. For two normal modes of vibration of a well-defined extended object, oscillator strengths should be comparable.
2. Collective modes should exhibit smooth properties as a function of nucleon number.
3. More subtly, figure 6.3 shows a systematic variation in the relative population of the 2^+ and 4^+ members of the gamma band.

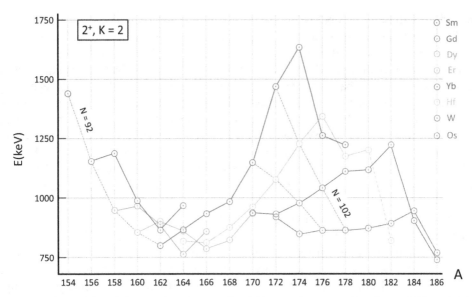

Figure 6.5. Systematics of the energies of gamma-band 2⁺ states in rare-earth region nuclei bounded by $N > 90$ and $Z < 76$. See text for details.

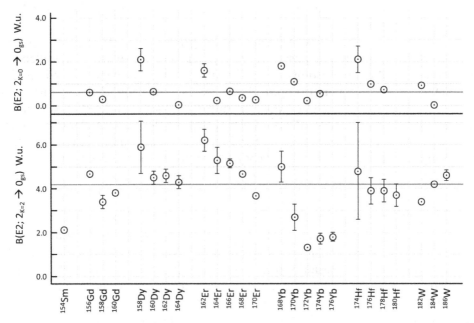

Figure 6.6. Systematics of $B(E2; 2^+ \rightarrow 0_1^+)$ for (a) the 2⁺ member of the lowest excited $K = 0$ band and (b) the 2⁺ member of the $K = 2$ gamma band, in rare-earth region nuclei bounded by $N > 90$ and $Z < 76$. Red lines indicate approximate median values for the $B(E2)$ strengths observed in these nuclei, i.e. ∼0.6 W.u. for the lowest $K = 0$ bands and ∼4 W.u. for the $K = 2$ bands. See text for details.

We address these points below.

The manifestation of excited $K = 0$ bands in deformed nuclei is complex: there are multiple $K = 0$ bands, already below an excitation of 2 MeV (cf figure 6.2); thus, besides any candidate beta band, what is the origin of the other $K = 0$ bands, and which one is the beta band? Spectroscopic data are very incomplete, both regarding properties of identified excited 0^+ states and regarding any confidence in a complete view, i.e., have all such states and their associated rotational bands, below 2 MeV, been identified? Above 2 MeV, there is no confidence in spectroscopic completeness for any deformed nucleus. In the entire rare earth region, only a few 2^+ members of $K = 0$ bands show $E2$ decay strength to the ground state exceeding 1 W.u., as depicted in figure 6.6. We address the effect of band mixing on observed $E2$ strengths later.

Collective modes in deformed nuclei should exhibit similar rotational properties. Indeed, one can assess this for the $K = 2$ bands depicted in figure 6.3 by comparing the 4^+-2^+ energy spacing with the ground-state band, gsb, e.g., for ^{160}Dy—190 keV ($K = 2$ band), 197 keV (gsb). Generally, for all the $K = 2$ bands shown in figure 6.3, the $K = 2$ band energy spacing is slightly smaller than the gsb energy spacing. However, for excited $K = 0$ bands in deformed rare-earth nuclei there are major differences. This can be seen in figure 6.2 where for ^{168}Er, the 2^+-0^+ energy spacings for $K = 0$ bands are—80 keV (gsb, $K = 0_1$), 59 keV ($K = 0_2$), 71 keV ($K = 0_3$), 59 keV ($K = 0_4$). This suggests shape coexistence, but to our knowledge has never been discussed (only noted [12]).

The peculiar trend in relative population of gamma band 2^+ and 4^+ states by inelastic deuteron scattering was already noted in the original work presented for dysprosium and erbium isotopes [3, 4], with similar behaviour noted in the gadolinium and ytterbium isotopes [2, 5]. It was conjectured to be due to a hexadecapole degree of freedom leading to direct population of the 4^+, $K = 2$ states, which results in interference effects with two-step, $E2$ population. To our knowledge, this universal feature of gamma bands has never been considered in collective modelling; yet there has been detailed characterization of this interference effect by a few experimental groups, and we give details shortly.

6.2 Octupole vibrations

Figure 6.3 shows population of 3^- states observed by inelastic scattering of deuterons. There is no evident systematic pattern. From detailed spectroscopy[2], K quantum numbers have been assigned to many of these states and are shown. The population of 3^- states by inelastic scattering has a natural interpretation in terms of collective octupole excitation. Figure 6.7 depicts the expected pattern of collective octupole excitation in a liquid drop model of a deformed nucleus, notably that four

---

[2] To be confident of the assignment of a K quantum number to a 3^- state, there must be detailed spectroscopic information to characterize all other nearby negative-parity states. Thus, for $K^\pi = 0^-$, nearby there must be only a lower 1^- state; but caution is needed because there are cases where this occurs and just above the 3^- state is a 2^- state, which indicates $K^\pi = 1^-$. The irregular ordering of the spins of the band members is due to mixing between states with the same spin-parity occurring in bands that are close in energy and differing by $\Delta K = 1$.

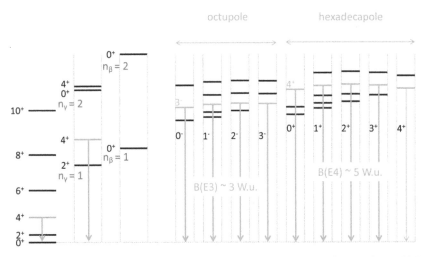

Figure 6.7. A multi-band view of excitations expected for a liquid-drop model of the nucleus. This includes multi-phonon beta and gamma vibrations, octupole vibrations and hexadecapole vibrations.

bands should be observed with $K^\pi = 0^-$, 1^-, 2^-, 3^-, not necessarily in this order by excitation energy. The states observed in the rare earth region that match this expectation are shown in figures 6.8(a)–(d). The first feature that should be noted is by far the most cases observed possess $K^\pi = 2^-$: a simple liquid drop model for understanding such states would predict the occurrence of equal numbers of $K^\pi = 0^-$, $K^\pi = 1^-$, $K^\pi = 2^-$, and $K^\pi = 3^-$ bands at low energy, distributed across the mass numbers presented. The second feature is that, comparing detailed spectroscopic information such as available for ^{168}Er and depicted in figure 6.2, many bands containing 3^- states are not observably populated, cf figure 6.1. We present a simple interpretation below.

A collective mode of the nucleus such as 'quadrupole' or 'octupole', whether permanent deformation or vibration, is most simply interpreted as a shape degree of freedom of a liquid drop. If this fails, a natural next step is to consider that, in some way, a subset of nucleons is selectively involved. Nucleons occupy orbitals and so it is natural to look at specific orbital combinations for the answer. Orbitals possess three basic ingredients: isospin, spin and spatial degrees of freedom. The simplest step is to attribute the observed selectivity of octupole collectivity to the spatial degree of freedom of orbitals. Thus, a selectivity that does not favour neutrons over protons and does not involve nucleon intrinsic-spin changes is indicated. This reduces to pairs of protons or pairs of neutrons in Nilsson configurations that originate in shell model configurations that differ in orbital angular momentum by 3, with the same spin–orbit alignments, that have opposite parity. These restrictions, when imposed on Nilsson diagrams for protons and neutrons, are depicted in figures 6.9(a) and (b). The observed dominance of $K^\pi = 2^-$ becomes evident: many pair combinations that are close in energy possess $K^\pi = 2^-$.

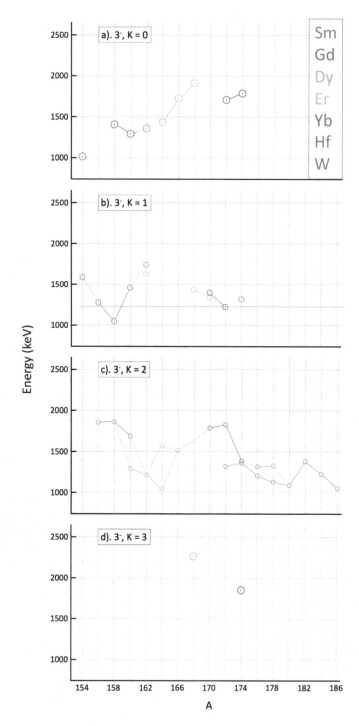

Figure 6.8. Systematics of excitation energies of collective 3⁻ states for nuclei in the rare earth region according to K quantum number assignments: (a) $K = 0$, (b) $K = 1$, (c) $K = 2$, and (d) $K = 3$. Note the suppressed energy zero. Details are discussed in the text.

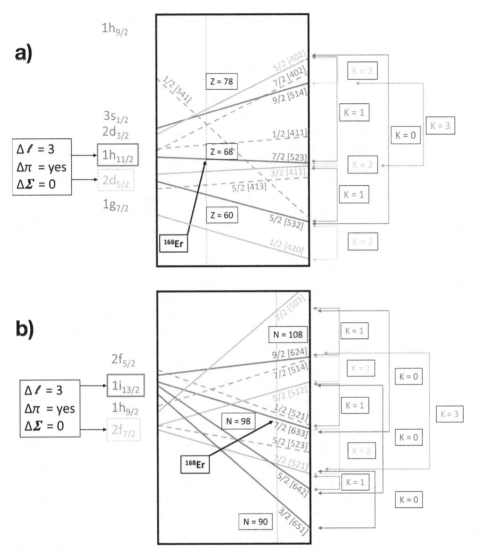

Figure 6.9. (a and b) Schematic view of Nilsson diagrams for the rare-earth region, (a) for protons and (b) for neutrons, illustrating the role of configurations that give rise to candidate collective octupole bands with $K = 0, 1, 2, 3$.

Deeper issues remain in connection with the above simple view: are the configurations identified in figures 6.9(a) and (b) observed to be close in energy in odd-mass nuclei adjacent to the K band observations? Are the configurations unmixed or mixed? Answering these questions requires detailed spectroscopy that identifies Nilsson configurations. Systematic features of Nilsson configurations will be handled in a future volume in the series.

6.3 Hexadecapole degrees of freedom and a unified view of collectivity

The remaining systematic observations emerging from inelastic scattering studies of deformed nuclei are well described by a collective hexadecapole degree of freedom. Strong direct $E4$ population of the 4_1^+ states in deformed nuclei is observed systematically via inelastic scattering and this mandates that nuclei possess either hexadecapole vibrations or hexadecapole deformation. This interpretation directly parallels interpretation of collective quadrupole degrees of freedom in nuclei via strong $E2$ population of 2_1^+ states. The distinction between multipole vibration and multipole deformation demands detailed spectroscopy: for $E2$ collectivity, a spectroscopic electric quadrupole moment of a state provides the answer, i.e. the determination of a diagonal $E2$ matrix element. (This distinction cannot be made for octupole collectivity because diagonal $E3$ matrix elements are always zero–from symmetry, any diagonal matrix element of an odd-rank spherical tensor is zero. The inference of octupole deformation in nuclei must be made via patterns of excitation and such information is too limited to arrive at a definitive answer.) Detailed patterns of excitation are also the procedure by which the Cd isotopes have been argued to possess quadrupole deformation, not quadrupole vibrations, as presented in section 5.2. Here, we look at the evidence for hexadecapole collectivity in nuclei; and we use the methods employed to make some more general remarks about elucidation of low-energy collective degrees of freedom in nuclei.

The spectroscopic evidence for hexadecapole degrees of freedom in nuclei is based on two spectroscopic views of the nucleus: inelastic scattering of charged particles and Coulomb excitation. There is an overlap between the two spectroscopic techniques that needs clarification of some subtle details, as follows.

Coulomb excitation always occurs when a charged particle interacts with a nucleus; for a charged particle that is hadronic (proton, deuteron, alpha particle, heavy ion, pion, ...), as opposed to leptonic (electron, muon, ...), excitation via the strong force also occurs if the charged particle approaches to within the interaction distance of this force (defined shortly). There is interference between the strong interaction and the Coulomb interaction. The Coulomb interaction probes charge distribution and dynamics in the nucleus. The strong interaction probes nucleon distribution and dynamics in the nucleus, i.e., the mass distribution. Also, depending on the hadronic probe, the relative involvement of protons and neutrons in an excitation mode may be revealed (inelastic scattering of π^+ and π^- beams is notably sensitive to this).

Figure 6.10 illustrates the way in which interference effects occur. Hexadecapole degrees of freedom are manifested through $E4$ multipole matrix elements, whether electromagnetic or mass. To distinguish between the two, we use '$E2$', 'β_2', '$E4$', and 'β_4' terminology. Added to these competing degrees of freedom, two-step, three step, $\cdots$ processes are manifested in direct competition for population of states with spin higher than 2 because of the dominance of $E2$ collectivity ($E2$ matrix elements) in nuclei.

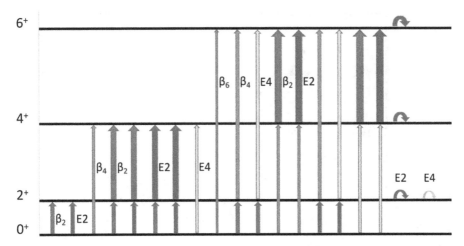

Figure 6.10. Schematic view of excitation of a rotational band by various combinations of inelastic scattering, β_2, β_4, β_6 and Coulomb excitation, $E2$, $E4$. Some Coulomb reorientation (diagonal) matrix elements are shown. Arrow widths qualitatively reflect spin dependence of symmetric top matrix elements. See text for more details.

Figure 6.11 shows how the possible excitation modes depicted in figure 6.10 are manifested in observations. The details reside in the angles with respect to the beam at which the particles are observed: these are referred to as scattering phase shifts or so-called 'Blair phase shifts'. Essentially, for each step involved in the scattering process (for an individual particle this can involve one-step, two-step, three-step, $\cdots$ processes), a phase shift occurs: think of the particle approaching the scattering target nucleus—it can be 'accelerated' (attracted) or 'retarded' (repelled)—this is manifested in its phase, expressed in terms of a plane-wave function or a spherical wave function. Further, this acceleration/retardation will occur differentially depending on the distribution of charge and mass in the target nucleus. Yet further, these processes depend on how deeply the beam probe penetrates the target nucleus. Figure 6.11 shows that the detection of scattered particles exhibits a $180°$ phase shift for population of the 2_1^+ state relative to the 0_1^+ state, i.e., one-step inelastic scattering versus elastic scattering. This process is dominated by the matrix element $\langle 2_1^+|E2|0_1^+\rangle$, but two-step processes such as $\langle 2_1^+|E2|2_2^+\rangle\langle 2_2^+|E2|0_1^+\rangle$ are possible—just improbable (so-called 'reorientation' effects are defined below). Again, figure 6.11 shows that the detection of scattered particles exhibits a $180°$ phase shift for population of the 4_1^+ state relative to the 2_1^+ state. The present interest is in the processes $\langle 4_1^+|E4|0_1^+\rangle$ versus $\langle 4_1^+|E2|2_1^+\rangle\langle 2_1^+|E2|0_1^+\rangle$, i.e. a one-step process versus a two-step process. We elaborate on this in the following.

For a purely coulombic interaction, the strong interaction effects play no role. This is guaranteed for electron scattering; for hadron scattering, the surfaces of the projectile and target must remain at least 5 fm apart (known as 'Cline's safe energy rule'). The measurement task is to analyze the scattered beam particles as a function of energy and angle with respect to the beam and to analyze the gamma-ray yields observed in association with each excited state (using gamma-ray spectroscopy). The

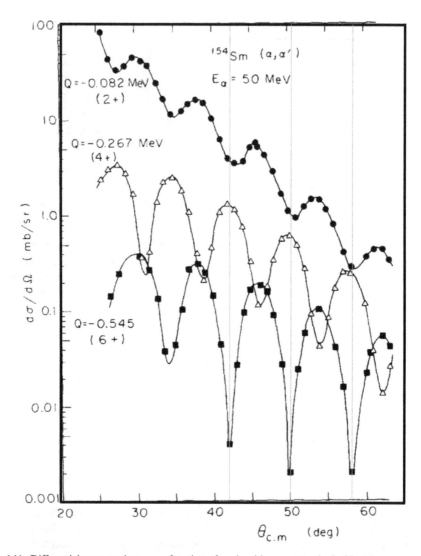

Figure 6.11. Differential cross sections, as a function of angle with respect to the incident beam expressed in the centre-of-mass system, for inelastic scattering of alpha particles on the deformed nucleus ^{154}Sm, for the spin-2, -4, and -6 members of the ground-state band. Note the $\sim 180°$ phase shifts between the patterns for each successive member of the band. Details are discussed in the text. Reprinted from [13], copyright (1967) with permission from Elsevier.

data analysis task for Coulomb excitation is deconvolution of the excitation paths, by which the state is populated, in terms of products of electromagnetic transition matrix elements. Thus, the 4_1^+ state may be populated by $\langle 4_1^+|E2|2_1^+\rangle\langle 2_1^+|E2|0_1^+\rangle$ or by $\langle 4_1^+|E4|0_1^+\rangle$. The process $\langle 4_1^+|E2|2_2^+\rangle\langle 2_2^+|E2|0_1^+\rangle$ is possible, but less likely because $\langle 2_2^+|E2|0_1^+\rangle$ is always much smaller than $\langle 2_1^+|E2|0_1^+\rangle$. The amplitudes for all probable paths are added to arrive at the total transition amplitude for population of the excited state. The phases of these composite amplitudes may not all be positive, which results in destructive interference with consequent attenuation of the population of the excited state. Different excitation paths will result in different orientations of the spin of the final state. The angular distribution (with respect to the beam direction) of emitted gamma rays is dictated by the orientation of the spin of the decaying state. One major effect that can occur is an excitation step followed by a 'reorientation' step. In its most familiar manifestation this would be $\langle 2_1^+|E2|2_1^+\rangle\langle 2_1^+|E2|0_1^+\rangle$ in studies of Coulomb excitation of 2_1^+ states; for population of the 4_1^+ states, $\langle 4_1^+|E2|4_1^+\rangle\langle 4_1^+|E4|0_1^+\rangle$ or $\langle 4_1^+|E4|4_1^+\rangle\langle 4_1^+|E4|0_1^+\rangle$ are possible. Such information is the way in which quadrupole moments (hexadecapole moments) can be extracted from Coulomb excitation measurements. This technique is referred to as 'Coulomb reorientation'. Figure 6.10 illustrates some of the pure Coulomb excitation paths that can occur in a multi-step Coulomb excitation measurement.

By a similar sequence of contributions, when Coulomb and strong interaction matrix elements are involved, composite amplitudes for excitation can be catalogued and used to analyze the spectroscopic signal for the excited state population: gamma rays emitted or inelastically scattered beam particle. Many issues arise in the quantification of matrix elements for strong interaction processes that populate collective states in nuclei. Thus, a discussion of so-called optical models for interaction of beam particles with nuclei is critical. The beam particle may be a proton, a deuteron, an alpha particle, or a heavier nucleus, often termed a 'heavy ion'. These are issues that are fundamental to nuclear structure study, but also embrace nuclear reaction theory. Issues such as the shape of the optical model potential (which leads, via this model to the nuclear shape), depth of penetration of the projectile into the target nucleus (recall that nuclei have diffuse surface densities), even spin–orbit interaction when the projectile has spin (and polarization—beams can be polarized), come into consideration. These lie beyond the present level of discussion; but we provide a few defining terms and a limited set of references.

Figure 6.12 illustrates the observational basis for inferring that hexadecapole matrix elements play a role in the excitation of the 4_1^+ and 6_1^+ states in the ground-state band of ^{154}Sm. This is manifested in the angular phase shifts for the respective populations of the 4_1^+ and 6_1^+ states and the details are defined in the figure and caption. The processes can be factored into population of the 4_1^+ state by the two-step process, $\langle 4_1^+|E2|2_1^+\rangle\langle 2_1^+|E2|0_1^+\rangle$ which is dominant because the matrix elements are both large, and by the one-step process $\langle 4_1^+|E4|0_1^+\rangle$. For population of the 6_1^+ state, the three-step process, $\langle 6_1^+|E2|4_1^+\rangle\langle 4_1^+|E2|2_1^+\rangle\langle 2_1^+|E2|0_1^+\rangle$ is dominant and there is competition from the two-step processes, $\langle 6_1^+|E4|2_1^+\rangle\langle 2_1^+|E2|0_1^+\rangle$ and $\langle 6_1^+|E2|4_1^+\rangle\langle 4_1^+|E4|0_1^+\rangle$. The one-step process, $\langle 6_1^+|E6|0_1^+\rangle$ can also be explored and would give an indication of tetrahexacontapole (2^6) degrees of freedom but is not pursued here.

Figure 6.13 shows the systematic features of $E4$ electromagnetic matrix elements/ optical model hexadecapole shape parameters for nuclei in the rare earth region converted to hexadecapole moments. We note that this is an interpretation of the matrix elements as reflecting permanent deformed shape components of these nuclei, not hexadecapole vibrations. Similar behaviour is observed in deformed nuclei in the

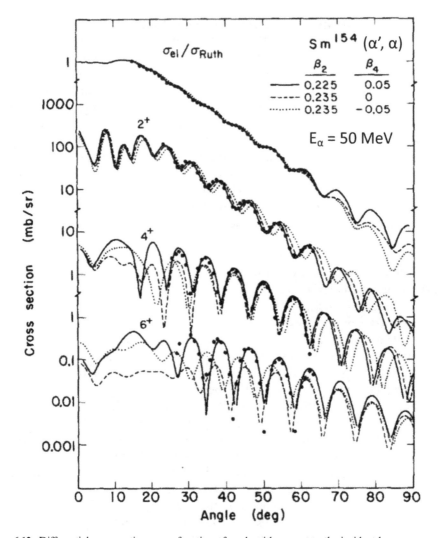

Figure 6.12. Differential cross sections, as a function of angle with respect to the incident beam expressed in the centre-of-mass system, for elastic scattering of 50 MeV alpha particles from ground-state band members of ^{154}Sm. The angular phase shifts reveal beyond the basic 180-degree shifts between successive band members that there are one-step $E4$ excitations that interfere with the two-step $E2 \times E2$ excitations: these shifts correspond to a hexadecapole shape component in the scattering potential with $\beta_4 \sim +0.05$, where the radius of the potential is expressed as $R = R_0(1 + \beta_2 Y_{20} + \beta_4 Y_{40})$ and $\beta_2 \sim +0.27$; shifts that would correspond to $\beta_4 = -0.05$ and 0.0 are also shown. Reprinted from [14], copyright (1968) with permission from Elsevier.

actinide region and supporting data are presented in table 6.1. Spectroscopic hexadecapole moments are observables for nuclei but are much harder to quantify than quadrupole moments.

The $E4$ matrix elements in the actinide region convert to $B(E4)$ values, some >100 W.u. If this is interpreted in terms of permanently deformed shapes, figure 6.14 illustrates a geometric view of these nuclei. The interesting and critical issue of the

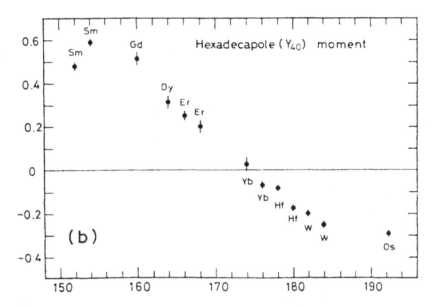

Figure 6.13. Systematics of Y_{40} shape components of the scattering potential for polarized protons across the rare earth region. Reprinted figure with permission from [15]. Copyright 1987 by the American Physical Society.

sign of the $E4$ matrix element arises and this has been directly resolved experimentally in a few cases: an example of the experimental evidence is depicted in figure 6.15. This has implications for deformed mean fields and the Nilsson model.

Figures 6.16(a) and (b) show evidence that gamma bands in nuclei in the deformed rare-earth region have a significant hexadecapole collective contribution in the form of a Y_{42} moment. A map of Y_{42} moments in the rare earth region is presented in figure 6.17. Again, this may be vibrational, or it may involve a permanent deformation. If it is a permanent deformation, gamma bands would appear to be a manifestation of a permanent axially asymmetric deformation. Whether deformational or vibrational, this manifestation necessitates a completely new approach to the Bohr model and its gamma vibrations. We take a deeper look at this below.

If the gamma bands in deformed nuclei are based on the classical Bohr model vibration, then two-phonon gamma bands should be observed. There will be two bands, one with $K = 4$ and one with $K = 0$ (recall that K is a directional quantization and does not have vectorial character). Many $K = 4$ bands are known in the rare earth region, but there is a lack of consensus on interpretation. Notably, many of these $K = 4$ bands that have possible enhanced $E2$ decay strength to the $K = 2$ gamma bands, are strongly populated in one-nucleon transfer reactions (some examples are presented below). A two-phonon structure cannot exhibit such behaviour. With the role of Y_{42} degrees of freedom in gamma bands, it would appear worthwhile to explore the role of Y_{44} degrees of freedom in $K = 4$ bands.

Table 6.1. A tabulation of $E4$ electromagnetic excitation strengths associated with the ground states of deformed actinide nuclei, for which such data are available (corresponding $E2$ transition strengths are also given). The transition strengths are expressed both in e^2b^2 ($E2$), e^2b^4 ($E4$) and in Weisskopf units, W.u. Note that the terminology used shows, e.g. $E2$ excitation of 2_1^+ states as $B(E2; 0 \to 2)$ where $B(E2)$ values are related to matrix elements as $B(E2); I_i \to I_f) = \langle I_f|E2| \rangle I_i^2/(2I_i + 1)$; thus, expressing $E2$ transition strengths between ground states and first excited 2^+ states differs between $0 \to 2$ and $2 \to 0$ by a factor of 5. The numbers quoted are for the $2 \to 0$ directions for the values given in W.u. and for the $0 \to 2$ directions for the values given in e^2b^2. The relationship between e^2b^2 and $E2$ W.u. is 1 W.u $(E2) = 5.940 \times 10^{-6}A^{4/3}$ e^2b^2; the relationship between e^2b^4 and $E4$ W.u. is 1 W.u. $(E4) = 6.285 \times 10^{-10}A^{8/3}$ e^2b^4. The table is based on one appearing in [16].

Nucleus	$B(E2; 0 \to 2)$ (e^2b^2)	$B(E2; 2 \to 0)$ (W.u.)	$B(E4; 0 \to 4)$ (e^2b^4)	$B(E4; 0 \to 4)$ (W.u.)	M_{04} $(E4)$ (eb^2)
^{230}Th	8.06^{11}	193	1.19^{32}	106	1.09^{15}
^{232}Th	9.21^{9}	217	1.48^{34}	129	1.22^{15}
^{234}U	10.90^{10}	256	1.96^{56}	167	1.40^{20}
^{236}U	11.60^{15}	268	1.69^{57}	140	1.30^{22}
^{238}U	12.30^{15}	281	0.69^{37}	56	0.83^{22}
^{238}Pu	12.63^{17}	288	1.90^{67}	154	1.38^{25}
^{240}Pu	13.33^{18}	301	1.31^{62}	104	1.15^{28}
^{242}Pu	13.47^{18}	301	$0.55^{+0.53}_{-0.41}$	43	0.74^{34}
^{244}Pu	13.61^{18}	301	$0.09^{+0.55}_{-0.09}$	7	$0.03^{+0.5}_{-0.8}$
^{244}Cm	14.58^{19}	322	$0.0^{+0.25}_{-0.0}$	0	$0.0^{+0.3}_{-0.5}$
^{246}Cm	14.94^{19}	326	$0.0^{+0.25}_{-0.0}$	0	$0.0^{+0.3}_{-0.5}$
^{248}Cm	14.99^{19}	324	$0.0^{+0.36}_{-0.0}$	0	$0.0^{+0.5}_{-0.6}$

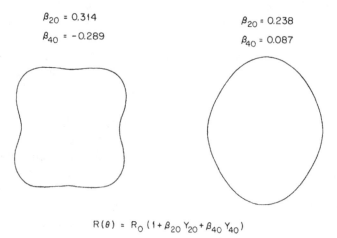

$$R(\theta) = R_0 (1 + \beta_{20} Y_{20} + \beta_{40} Y_{40})$$

Figure 6.14. A depiction of the difference between a hexadecapole shape, superimposed on quadrupole shape, for the deformation parameters shown. Reprinted figure with permission from [17], copyright 1973 by the American Physical Society.

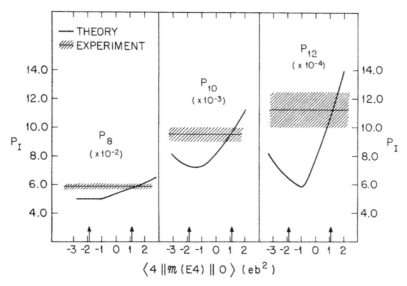

Figure 6.15. Observations that support a positive sign of $E4$ for the matrix elements (in ^{238}U). The quantities P_8, P_{10}, P_{12} are Coulomb excitation probabilities for the spin 8, 10, 12 states of the ground-state band. The solid lines show the variations of the Coulomb excitation probabilities versus the values of $\langle 4\|M(E4)\|0\rangle$ in eb^2, and the vertical arrows distinguish between the values -2 and $+1$ for this matrix element with a clear indication that it is $+1$ (earlier results were ambiguous). Reprinted figure with permission from [17], copyright 1973 by the American Physical Society.

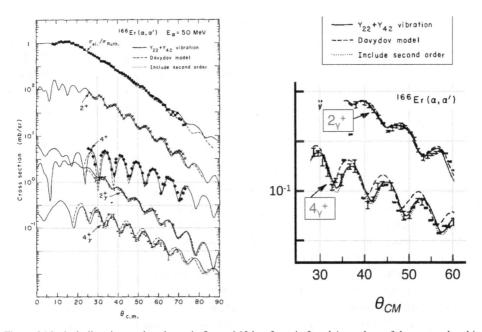

Figure 6.16. A similar view to that shown in figure 6.12 but for spin 2 and 4 members of the gamma band in ^{166}Er. The phase shifts indicate that the 4^+ member of the gamma band has a shape component described as a near equal mixture of Y_{22} and Y_{42} shape moments. Reprinted from [18], copyright (1969) with permission from Elsevier.

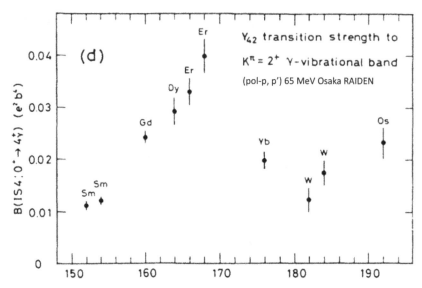

Figure 6.17. Systematics of Y_{42} shape components of the scattering potential for polarized protons inelastically scattered from some rare-earth nuclei. Reprinted figure with permission from [15], copyright 1987 by the American Physical Society.

There is no consensus on the possible manifestation of $K = 0$, two-phonon gamma vibrational bands.

The issues of collectivity associated with $K = 2$, gamma bands and the hexadecapole degree of freedom come to a well-studied focus in the osmium isotopes, ^{186,188,190,192}Os. These isotopes are discussed as candidate nuclei for axially asymmetric rotation in section 2.5. It was noted that the model used, the GTRM, while showing a good description of $E2$ matrix elements in the spin-0, 2 subspace, fails when applied to the spin-0, 2, 4 subspace in these nuclei; but no details were given. In the following we give details of why the model does not work for states with spin 4 in these osmium isotopes.

The isotopes ^{186,188,190,192}Os possess three 4$^+$ states below 1400 keV, cf figure 2.2. The $E2$ properties of these states are shown in table 2.3. From the perspective of the GTRM, the energy of the 4_3^+ state does not match expectations, cf figure 2.3; but it does more nearly match expectations of a harmonic gamma-vibrational model ($E_{K=4} = 2 \times E_{K=2}$). However, in all these isotopes, the $K = 4$ band head is strongly populated in one-proton transfer reactions, and this is shown for ^{188}Os in figure 6.18. Further, inelastic scattering directly populates these $K = 4$ band heads strongly as shown for ^{188}Os in figure 6.19. Thus, they are neither candidate $K = 4$ bands for the GTRM nor for the gamma vibrational model. However, by invoking a hexadecapole component to the shape of these isotopes, a consistent picture emerges. Some details are given below.

A consistent picture of collective properties of ^{186,188,190,192}Os is achieved by proposing that there exists a fourth 4$^+$ state and an associated $K = 4$ band. This band mixes with the $K = 4$ band of the GTRM model. Evidence for this comes from a sum rule for $E2$ transitions in a deformed nucleus; this is shown in figure 6.20. There is

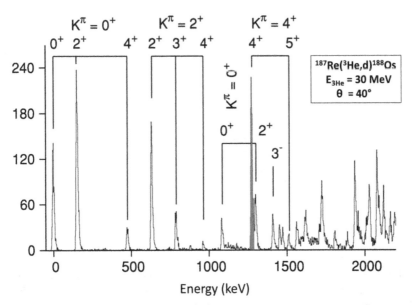

Figure 6.18. Spectrum of deuterons from the one-proton transfer reaction ^{187}Re(^{3}He,d)^{188}Os which reveals that the $K = 4$ band based on the 4$^+$ state at 1279 keV (cf figure 2.2) possesses a broken-pair (non-collective) component. Reprinted figure with permission from [19], copyright 2010 by the American Physical Society.

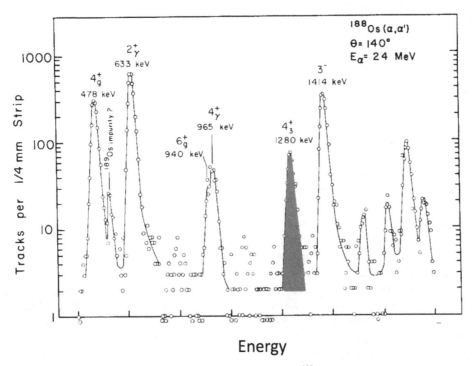

Figure 6.19. Spectrum of alpha particles from inelastic scattering on ^{188}Os which reveals that the $K = 4$ band based on the 4$^+$ state at 1279 keV (cf figure 2.2) possesses a collective hexadecapole component. Reprinted from [20], copyright (1978) with permission from Elsevier.

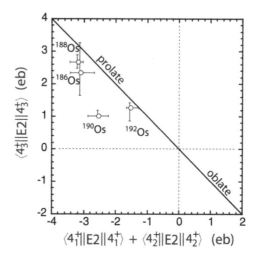

Figure 6.20. Sum rule for diagonal $E2$ matrix elements for 4+ states in ^{186,188,190,192}Os, showing that not all the collective $E2$ strength in association with the 4+ states has been observed. Reprinted figure with permission from [22], copyright 2013 by the American Physical Society.

further evidence from the one-proton transfer reaction spectroscopy: states with probable $K = 4$ are populated at ~2.5 MeV [21], and references therein.

One feature evident in rare-earth region nuclei associated with a hexadecapole degree of freedom remains to be considered. In some nuclei, a $K = 3$ band is observed with a 4+ band member which is strongly populated by inelastic scattering. Figures 6.1 and 6.2 identify such a structure in ^{168}Er at 1737 keV. This matches the $K^\pi = 3^+$ band possibility shown in figure 6.7. But it opens the question: 'where are the $K^\pi = 0^+$, 1+, 2+ and 4+ bands?' One approach is to follow the interpretation made for 3$^-$ collective, octupole states and consult a Nilsson diagram for possible $\Delta l = 4$, $\Delta \pi = $ no, $\Delta \Sigma = 0$ parentage combinations, similar to the procedure shown in figures 6.9(a) and (b). Use of Nilsson diagrams to identify such configurations in comparison with data will be handled in a future volume in the series.

The conclusion from the foregoing details of hexadecapole degrees of freedom in deformed nuclei is that they are present with a systematic pattern. Further, they are manifested as an integral component of gamma bands, i.e. gamma bands cannot be viewed as 'Y_{22}' multipole degrees of freedom in nuclei, they are '$Y_{22} + Y_{42}$' degrees of freedom. Beyond this stated view, experimental investigation is needed.

Multi-phonon beta vibrational bands, all of which have $K = 0$, have been discussed in the literature; but the very weak $E2$ transition strength manifested in the one-phonon candidate bands suggests that an unequivocal proof of the existence of such structure will not be easy. Band mixing, such as manifested in ^{152}Sm, cf figure 4.9, will complicate such searches: note that the inter-band strength arises entirely from mixing.

We make some closing comments on the occurrence of low-energy vibrational degrees of freedom in nuclei, based on the view that emerges in this chapter. First, any low-energy vibrational collectivity is weak. Probably this strength is associated with specific proton and neutron orbitals in the manner illustrated for octupole collectivity in figures 6.10(a) and (b). Searches for collective strength are effectively made following the procedures presented for hexadecapole strength using inelastic scattering and multi-step Coulomb excitation.

6.4 Exercises

6-1. Figure 6.2 shows the very extensive level scheme for ^{168}Er organized into various rotational bands with suggested K^π assignment. Using data found in ENSDF, as far as possible, carry out a similar classification for excited states in
(a) ^{166}Er;
(b) ^{170}Yb;

(Limit the initial classification to: positive parity, first $K = 2$ band, first two $K = 0$ bands, first $K = 4$ band; negative parity, first $K = 0, 1, 2, 3$ bands; first $K^\pi = 1^+$, 3^+, 5^+, 4^-, 5^- bands.)

Having carried out such classification into rotational bands, it is possible to explore differences on changing the number of neutrons or protons:
(c) Identify the similarities in the structure of ^{166}Er and ^{168}Er which are neighbouring even–even isotopes;
(d) Identify the similarities in the structure of ^{170}Yb and ^{168}Er which are isotones i.e. have the same number of neutrons, viz. $N = 100$.

6-2. In a manner similar to exercise 6-1: using data found in ENSDF, as far as possible, carry out a classification of excited states in:
(a) ^{234}U;
(b) ^{236}U;
(c) Identify the similarities in the structure of the neighbouring even–even isotopes, ^{234}U and ^{236}U.

6-3. Figure 6.5 presents systematics of the energies of gamma-band 2^+ states in rare earth nuclei. Using ENSDF, compile, as far as possible, similar data for the actinide isotopes.

6-4. Figure 6.6 presents systematics of $B(E2; 2^+ \rightarrow 0_1^+)$ values for the 2^+ members of the lowest excited $K = 0$ band and the $K = 2$ gamma band in rare earth nuclei. Using ENSDF, compile, as far as possible, similar data for the actinide isotopes.

6-5. Figure 6.8 presents systematics of excitation energies of collective 3^- states in rare earth nuclei. Using ENSDF, compile, as far as possible, similar data for the actinide isotopes.

6-6. A long-standing candidate for a beta vibration was the first excited 0^+ state in ^{152}Sm at 685 keV, cf figure 6.21. This nucleus, at the location for strong onset of deformation, between ^{150}Sm ($N = 88$) and ^{154}Sm ($N = 92$), appeared to be the textbook case of a beta vibration. The sudden onset of deformation suggested that at $N = 90$ nuclei might be 'soft' with respect to deformability in the parameter β and so 'β' vibrations would naturally be observed. The $B(E2)$ for the 563 keV, $0_2^+ \rightarrow 2_1^+$ transition, with a value of

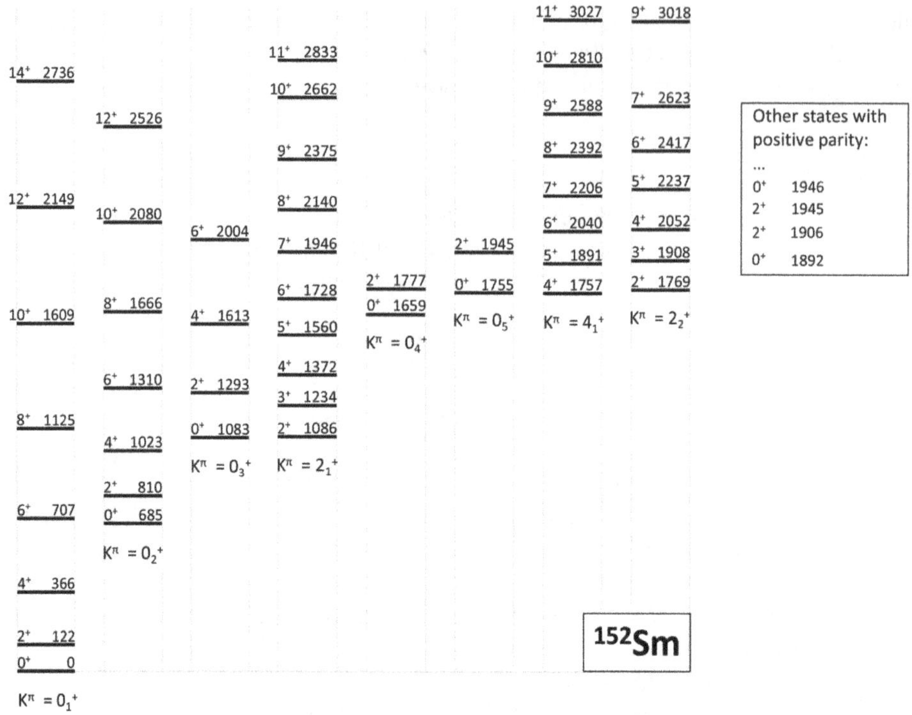

Figure 6.21. The low-lying positive-parity states in ^{152}Sm organized into rotational bands with differing values for the K quantum number.

Positive-parity states in ^{152}Sm (energies in keV), organized into rotational bands:

$K^\pi = 0_1^+$: 0^+ 0, 2^+ 122, 4^+ 366, 6^+ 707, 8^+ 1125, 10^+ 1609, 12^+ 2149, 14^+ 2736

$K^\pi = 0_2^+$: 0^+ 685, 2^+ 810, 4^+ 1023, 6^+ 1310, 8^+ 1666, 10^+ 2080, 12^+ 2526

$K^\pi = 0_3^+$: 0^+ 1083, 2^+ 1293, 4^+ 1613, 6^+ 2004

$K^\pi = 2_1^+$: 2^+ 1086, 3^+ 1234, 4^+ 1372, 5^+ 1560, 6^+ 1728, 7^+ 1946, 8^+ 2140, 9^+ 2375, 10^+ 2662, 11^+ 2833

$K^\pi = 0_4^+$: 0^+ 1659, 2^+ 1777

2^+ 1945

$K^\pi = 0_5^+$: 0^+ 1755, 5^+ 1891, 7^+ 2206, 9^+ 2588, 10^+ 2810, 11^+ 3027

$K^\pi = 4_1^+$: 4^+ 1757, 6^+ 2040, 8^+ 2392, 9^+ 3018

$K^\pi = 2_2^+$: 2^+ 1769, 3^+ 1908, 4^+ 2052, 5^+ 2237, 6^+ 2417, 7^+ 2623

Other states with positive parity:
...
0^+ 1946
2^+ 1945
2^+ 1906
0^+ 1892

^{152}**Sm**

---

33 W.u., appears to support this view. However, the shape coexistence view and the generation of this $E2$ strength entirely through mixing of configurations with different deformations make the beta-vibrational view questionable. We explore the apparent contradiction in this exercise.

(a) Using data in ENSDF, plot the excitation energies of the 2_1^+, 4_1^+, 6_1^+ and 8_1^+ states in ^{144,146,148,150,152,154,158,160}Sm.

A vibrational degree of freedom should exhibit a two-phonon excitation. With a one-phonon 'strength' of 33 W.u. de-exciting a 0^+ state at 685 keV, a two-phonon to one-phonon de-excitation from a state at about 1370 keV excitation (2×685) should exhibit a $B(E2)$ of about 66 W.u. (2×33). This should be a $K = 0$ band, cf figure 6.7, and the strength would be manifested in $0_{\beta\beta} \rightarrow 2_\beta$, $2_{\beta\beta} \rightarrow 0_\beta$ and $2_{\beta\beta} \rightarrow 2_\beta$ transitions. This has been investigated by multi-step Coulomb excitation.

Figure 6.22 shows de-excitation in ^{152}Sm, through the 0^+, 685 keV state, as observed by coincidence gating on the 563 keV gamma ray, following Coulomb excitation of a ^{152}Sm beam on a ^{208}Pb target. Besides the expected gamma rays from the rotational band built on the 685 keV state (note they are very strongly attenuated because the band drains 99.6% into the ground-state band at the 810 keV state due to the E^5 factor in radiative decay via $E2$ transitions) a very strong feeding is observed via

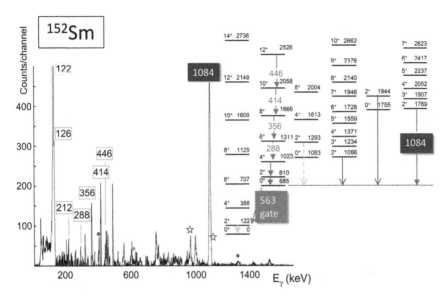

Figure 6.22. A coincidence spectrum gated on the 563 keV, $685(0^+_2) \rightarrow 122(2^+_1)$ transition to show the feeding of the 685 keV state following Coulomb excitation of a ^{152}Sm beam on a ^{208}Pb target. Note the very strong decay from the 1769 keV state by the 1084 keV transition, this is discussed further in the text. Also note that the band built on the 685 keV state shows only weakly in coincidence because it drains out 99.7% at the 810 keV level. The figure is provided courtesy of David Kulp.

a gamma ray of 1084 keV. A double coincidence gating on the 122 (122 → 0) and 689 (810 → 122) gamma rays establishes that this strength comes from a state at 1769 keV, via transitions 1084 (1769 → 685) and 959 (1769 → 810), as shown in figure 6.23. A natural interpretation would be that the 1769 keV state is a manifestation of the expected two-phonon beta vibration. This would be the spin-2 member of a $K = 0$ band, cf figure 6.7. Except that this spin-2 state is the head of a $K = 2$ band. The proof of this is detailed below.

(b) Using data in ENSDF, determine the fraction of the population of the beta band which, on reaching the 810 keV 2^+ band member, feeds the 685 keV 0^+ band member.

(c) Repeat the process in (ii) for the population of the 1023 keV 4^+ beta band member reaching the 810 keV 2^+ band member.

(d) The $0_{\beta\beta} \rightarrow 2_\beta$, $2_{\beta\beta} \rightarrow 0_\beta$ and $2_{\beta\beta} \rightarrow 2_\beta$ transition strengths will be impacted by Clebsch–Gordan coefficients, $\langle 0020|20 \rangle$, $\langle 2020|00 \rangle$, and $\langle 2020|20 \rangle$ respectively. What are the numerical values of these coefficients?

(e) Why is spin zero excluded for the 1769 keV state?

Proof of a K quantum number assignment to a state in a deformed nucleus must be done by indirect means. By association with other states, a spin-2 state can be inferred to have $K = 0$ if it lies above a spin-0 state at a similar energy to that observed in the ground-state band (in ^{152}Sm, this is 122 keV). In turn, (in ^{152}Sm) about 244 keV above it will be a spin-4

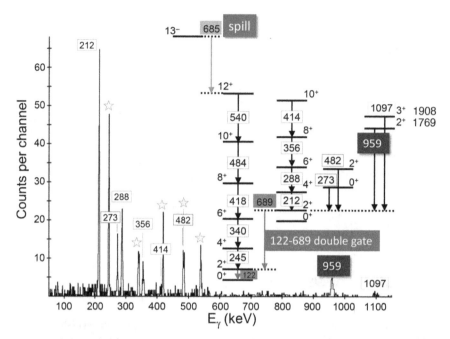

Figure 6.23. A coincidence spectrum double-gated on the 689 keV, $810(2_2^+) \rightarrow 122(2_1^+)$ and 122 keV, $122(2_1^+) \rightarrow 0(0_1^+)$ transitions to show the feeding of the 810 keV state following Coulomb excitation of a ^{152}Sm beam on a ^{208}Pb target. Note the very strong decay from the 1769 keV state by the 959 keV transition, this is discussed further in the text. Also note that the band built on the 685 keV state shows only weakly in coincidence because it drains out 92% at the spin-4 (1023 keV) level. There is also a 'spill' into the 689 keV gate due to a 685 keV transition which feeds the 12^+ member of the ground-state band: the resulting coincident lines are indicated by stars, without energy labels. The figure is similar to one appearing in [23].

state. If the spin-2 state has $K = 2$, there will not be a candidate spin-0 state below it, above it will be candidate spin-3, -4, -5 ···states. An immediate issue is whether or not all relevant states have been seen; excited 0^+ states in nuclei are notoriously difficult to populate. There is a second, more direct but subtle way to assign a K quantum number to a spin-2 state: there is an enormous difference between the Clebsch–Gordan coefficients $\langle 2020|40 \rangle$ and $\langle 222, -2|40 \rangle$. Thus, a spin-2 state with $K = 2$ only weakly decays to the spin-4 member of the ground-state band, whereas if it has $K = 0$ it predominantly decays to the spin-4 member of the ground-state band (remember to account for the E^5 factor in observed radiative decay strength).

(f) Using data for ^{152}Sm in ENSDF, what are the branching ratios to the spin-0, -2 and -4 members of the ground-state band for: (i) the 2^+ 810 keV state? (ii) the 2^+ 1086 keV state? (iii) the 2^+ 1769 keV state? (Note: the $1769 \rightarrow 366$ transition is not reported in ENSDF, i.e. it has not been seen, so it must be very weak.)

(g) If the 2^+ 1769 keV state was a $K = 0$ state, based on the decay branch $1769 \rightarrow 0$, what would be the decay branch $1769 \rightarrow 366$? (Remember the E^5 factor.)

A powerful spectroscopic technique that provides a comprehensive (near complete) view of excited states in a nucleus, for states up to a maximum spin of ~6, is $(n, n'\gamma)$ spectroscopy, i.e. γ-ray spectroscopy following inelastic scattering of neutrons. This is especially powerful when the neutron beam is (near) monoenergetic and collimated. Then by increasing the neutron energy in steps (e.g. 100 keV per step; a 'narrow' beam-energy spread is about ±50 keV), all low-spin excited states can be mapped up to the maximum available neutron beam energy. The threshold for appearance of a γ ray as a function of increasing neutron bombardment energy localizes the excitation of the state from which it originates. There is a distinct pattern of intensity for the population of an excited state as a function of increasing neutron bombardment energy, that depends on spin. Angular distributions of γ rays with respect to the beam direction also depend on spin. Yet further, lifetimes of excited states lead to characteristic energy shifts as a function of scattering angle due to Doppler effects resulting from each recoiling nucleus which is the source of the γ ray. Such data for ^{152}Sm$(n, n'\gamma)$ are shown in figures 6.24, 6.25 and 6.26.

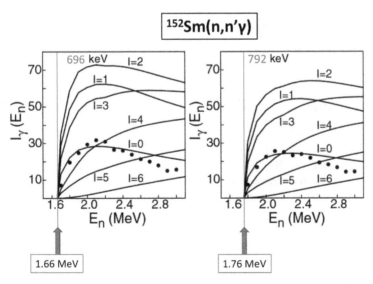

Figure 6.24. Excitation functions versus neutron bombarding energy for the 696 and 792 keV γ rays in a ^{152}Sm$(n,n'\gamma)$ study. These γ rays respectively de-excite the 0^+ excited states at 1659 and 1755 keV (they both feed the 963 keV state). The excitation thresholds of 1.66 and 1.76 MeV support the origins of these γ rays and the excitation patterns versus neutron bombarding energy support spin assignments of 0 to the 1659 and 1755 keV states. Reprinted figure with permission from [23], copyright 2008 by the American Physical Society.

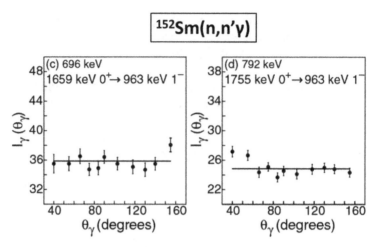

Figure 6.25. Angular distributions of the 696 and 792 keV γ rays with respect to the neutron beam direction in a ^{152}Sm$(n,n'\gamma)$ study. The 'flat' distribution further supports the assignment of spin 0 to the 1659 and 1755 keV states. Reprinted figure with permission from [23], copyright 2008 by the American Physical Society.

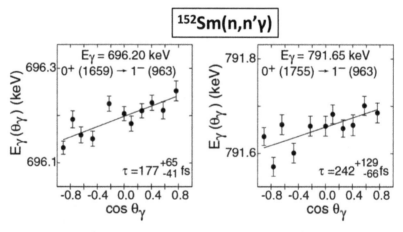

Figure 6.26. Doppler energy shifts of the 696 and 792 keV γ rays shown as a function of angle with respect to the neutron beam direction in a ^{152}Sm$(n,n'\gamma)$ study. These data establish lifetimes of 177 fs and 242 fs for the 1659 and 1755 keV states, respectively. Reprinted figure with permission from [23]. Copyright 2008 by the American Physical Society.

A View of Nuclear Data

Low-energy nuclear vibrations
10. Deformed nuclei: candidate vibrations

Data for deformed nuclei support vibrations at the one-phonon level.

Patterns suggest that these excitations depend on specific proton and neutron configurations.

Multipolarities (spin modes) with angular momentum 2, 3, and 4 are supported.

Tutorial 6.1 Deformed nuclei: candidate vibrations. The video can be downloaded from https://doi.org/10.1088/978-0-7503-5643-5.

A View of Nuclear Data

13. Octupole degrees of freedom in nuclei
Vibration? Deformation?

Collective octupole excitation in nuclei is summarized

Circumstantial evidence for octupole deformation is presented

Tutorial 6.2 Octupole degrees of freedom in nuclei. Vibration? Deformation?. The video can be downloaded from https://doi.org/10.1088/978-0-7503-5643-5.

A View of Nuclear Data

14. Hexadecapole degrees of freedom in nuclei
Deformation? Vibration?

Hexadecapole degrees of freedom in nuclei depend on inelastic scattering for identification.

Collective structures are revealed directly by intensities in scattered particle spectra

Collective structures are also revealed by interference effects when competing excitation paths are involved: these effects are observed in angular distributions with respect to the beam

It is important to recognize that inelastic electron scattering is governed purely by elect but inelastic scattering of (charged) hadrons involves both the em and str

Scattering data indicate that many nuclei possess a hexadecapole deformation "superimpos

Tutorial 6.3 Hexadecapole degrees of freedom in nuclei. Deformation? Vibration?. The video can be downloaded from https://doi.org/10.1088/978-0-7503-5643-5.

A View of Nuclear Data

Experimental probes
12. Inelastic scattering: hadrons

Inelastic scattering, particularly of deuterons and alpha particles (equal numbers of protons and neutrons), are shown to be an excellent probe of one-step and few-step excitation of nuclear collective modes.

Such data can "see" details of competing excitation modes, e.g. two-step E2 vs. one-step E4 excitation, through interference effects

Tutorial 6.4 Inelastic scattering: hadron probes. The video can be downloaded from https://doi.org/10.1088/978-0-7503-5643-5.

References

[1] Veje E, Elbek B, Herskind B and Olesen M C 1968 Inelastic deuteron scattering from ^{148}Sm, ^{150}Sm, ^{152}Sm and ^{154}Sm *Nucl. Phys.* A **109** 489–506

[2] Bloch R, Elbek B and Tjøm P J 1967 Collective vibrational states in even Gd nuclei *Nucl. Phys.* A **91** 576–92

[3] Grotdal T, Nybø K, Thorsteinsen T and Elbek B 1968 Collective vibrational states in even dysprosium nuclei *Nucl. Phys.* A **110** 385–99

[4] Tjøm P O and Elbek B 1968 Collective vibrational states in even erbium nuclei *Nucl. Phys.* A **107** 385–401

[5] Burke D G and Elbek B 1967 A study of energy levels in even–even ytterbium isosoptes by means of (d,p), (d,t) and (d,d') reactions *Mat. Fys. Medd. Dan. Vid. Selsk.* **36** (6)

[6] Günther C, Kleinheinz P, Casten R F and Elbek B 1971 Vibrational states in even wolfram nuclei *Nucl. Phys.* A **172** 273–97

[7] Walker P M 1983 Deformation aligned states *Phys. Scr.* **T5** 29–35

[8] Jenkins D G and Wood J L 2021 *Nuclear Data: A Primer* (Bristol: IOP Publishing) https://iopscience.iop.org/book/mono/978-0-7503-2674-2

[9] Rowe D J and Wood J L 2010 *Fundamentals of Nuclear Models: Foundational Models* (Singapore: World Scientific)

[10] Bohr A 1952 The coupling of nuclear surface oscillations to the motion of individual nucleons *Mat. Fys. Medd. Dan. Vid. Selsk.* **26** (14)

[11] Bohr A and Mottelson B 1953 Collective and individual-particle aspects of nuclear structure *Mat. Fys. Medd. Dan. Vid. Selsk.* **27** (16)

[12] Heyde K and Wood J L 2011 Shape coexistence in atomic nuclei *Rev. Mod. Phys.* **83** 1467

[13] Harvey B G, Hendrie D L, Jarvis O N, Mahoney J and Valentin J 1967 Multiple nuclear excitation of rotational levels in ^{152}Sm and ^{154}Sm *Phys. Lett.* B **24** 43–6

[14] Hendrie D L, Glendenning N K, Harvey B G, Jarvis O N, Duhm H H, Saudinos J and Mahoney J 1968 Determination of Y_{40} and Y_{60} components in the shapes of rare earth nuclei *Phys. Lett.* B **26** 127–30

[15] Ichihara T, Sakaguchi H, Nakamura M, Yosoi M, Ieiri M, Takeuchi Y, Togawa H, Tsutsumi T and Kobayashi S 1987 Inelastic proton scattering exciting the γ-vibrational band in deformed nuclei ($152 \leqslant a \leqslant 192$) at 65 MeV and the systematics of the hexadecapole (Y_{42}) strength of the γ vibration *Phys. Rev.* C **36** 1754

[16] Bemis C E, McGowan F K, Ford J L C, Milner W T, Stelson P H and Robinson R L 1973 $E2$ and $E4$ transition moments and equilibrium deformations in the actinide nuclei *Phys. Rev.* C **8** 1466

[17] Eichler E, Johnson N R, Sayer R O, Hensley D C and Riedinger L L 1973 Sign of the hexadecapole moments of ^{232}Th and ^{238}U nuclei *Phys. Rev. Lett.* **30** 568

[18] Mackintosh R S 1969 The nature of the gamma band in erbium and dysprosium *Phys. Lett.* B **29** 629–31

[19] Philips A A *et al* 2010 Structure of the K $^\pi = 4^+$ bands in $^{186,\,188}$Os *Phys. Rev.* C **82** 034321

[20] Burke D G, Shahabuddin M A M and Boyd R N 1978 Hexadecapole vibrations in osmium isotopes *Phys. Lett.* B **78** 48–51

[21] Burke D G 1994 Hexadecapole-phonon versus double-γ-phonon interpretation for $K^\pi = 4^+$ bands in deformed even-even nuclei *Phys. Rev. Lett.* **73** 1899

[22] Allmond J M 2013 Simple correlations between electric quadrupole moments of atomic nuclei *Phys. Rev.* C **88** 041307(R)

[23] Kulp W D *et al* 2008 Search for intrinsic collective excitations in ^{152}Sm *Phys. Rev.* C **77** 061301(R)

IOP Publishing

Nuclear Data
A collective motion view
David Jenkins and John L Wood

Chapter 7

Epilogue

The material presented herein, which is aimed at providing a basic introduction to the collective structure behind nuclear data, reveals features that provide a strong guide to future work. The following discussion focusses on these features.

The over-riding message from nuclear data is that many nuclei conform to (near) axially symmetric rotor behaviour. Most notably the K quantum number is realized as a good quantum number and is well handled by the symmetric top model. The K quantum number is an emergent property of nuclei, i.e., it appears through the self-organization of nuclei as finite many-body systems. This validity of the K quantum number has a high impact on handling nuclear data. It enables reduction of many measured quantities, via the Wigner–Eckart theorem and Clebsch–Gordan coefficients, to yield intrinsic quadrupole moments, Q_0. This view of nuclear deformation through Q_0 values reveals a zeroth-order measure of nuclear deformation that shows a consistency at about the 2% level across all strongly deformed nuclei, both even and odd mass. For a few strongly deformed nuclei, high-precision tests of simple model mixing of the K quantum number are available which support a consistency at about the 1% level.

There is a simple model extension to test asymmetric top behaviour in nuclei, the general triaxial rotor model (GTRM). This can handle a range of observations pertaining to $E2$ matrix elements in deformed even–even nuclei, especially first- and second-excited 2^+ states; further, it provides a useful treatment of $\Delta K = 2$ band mixing.

Deformation appears widely, even in nuclei with spherical ground states. This is handled under the title of shape coexistence. It is possible that shape coexistence occurs in all nuclei; but much experimental work is needed to explore the extent of occurrence and the range of deformations realized in nuclei. Superdeformed bands lack detailed spectroscopic characterization; but suggest that such bands approach a near rigid-body limit for the symmetric top.

Vibrational collectivity at low energy in nuclei is poorly defined and is confined to a few configurations with no compelling evidence for multi-phonon states. Indeed, where vibrational behaviour had been suggested, e.g., quadrupole vibrations in the Cd isotopes, detailed spectroscopy supports K band structures, i.e. rotational behaviour.

We note that simple models and mixing of configurations with different deformations appear to be sufficient to describe many features in nuclear data. This is at odds with the modern trend into complex models with intricate tuning of many parameters. Indeed, it suggests that intricate and complex theories may be a wrong approach to understanding nuclear structure: nuclear structure is much simpler. However, it requires detailed spectroscopy to reveal the way in which simple model configurations mix.

The over-riding message from this view of data is that rotations emerge as a strongly collective mode which is independent of nucleon configurations in its primary manifestation. Vibrations, in contrast, are weakly collective modes that depend on specific nucleon configurations and therefore multi-phonon excitations are elusive or non-existent because of the Pauli exclusion principle.

This brief summary points to spherical and weakly deformed nuclei as the next step in assessing the structure behind nuclear data. These topics will be taken up in following volumes in the Series.

IOP Publishing

Nuclear Data
A collective motion view
David Jenkins and John L Wood

Appendix A

Derivation of commutator bracket relations for spin in a body-fixed frame

The commutator brackets for the components of angular momentum, e.g., for three Cartesian axes x, y, z in $(3, R)$, the space in which we live—the 'laboratory' frame, take the familiar form

$$[L_x, L_y] = i\hbar L_z, \tag{A.1}$$

with cyclic permutations of x, y, z. The commutator brackets for the components of nuclear spin in the body-frame of the nucleus—the 'intrinsic' frame, e.g., for three body-fixed axes 1, 2, 3 take the form

$$[L_1, L_2] = -i\hbar L_3 \tag{A.2}$$

with cyclic permutations of 1, 2, 3. Note the minus sign. The reason for this is not obvious. The following provides a concise derivation of equation (A.2) from equation (A.1).

The key to connecting equations (A.1) and (A.2) is the recognition that there is a rotational transformation between the intrinsic frame of the nucleus and the laboratory frame. Rotational transformations about an axis 'n' take the generic form

$$R_n(\phi_n) = \exp\left\{-\frac{i(\boldsymbol{n} \cdot \boldsymbol{L})\phi_n}{\hbar}\right\}, \tag{A.3}$$

where $\boldsymbol{n} \cdot \boldsymbol{L}$ is the component of angular momentum along the n-axis and the rotation angle is ϕ_n. Thus, one can build a language of successive rotational transformations and effect a rotation between any two frames of reference in a given space. To ensure that this language correctly describes rotations in $(3, R)$ it is important to recall that rotations about different axes do not commute, even classically. Thus, for rotations about the x- and y-axes, using $R_x(\phi_x) = \exp\{-iL_x\phi_x/\hbar\}$ and $R_y(\phi_y) = \exp\{-iL_y\phi_y/\hbar\}$, consider

$$R_y(\phi_y)R_x(\phi_x) = \exp\{-iL_y\phi_y/\hbar\} \exp\{-iL_x\phi_x/\hbar\} \qquad (A.4)$$

and

$$R_x(\phi_x)R_y(\phi_y) = \exp\{-iL_x\phi_x/\hbar\} \exp\{-iL_y\phi_y/\hbar\} \qquad (A.5)$$

i.e., rotation about the x-axis through an angle ϕ_x followed by rotation about the y-axis through an angle ϕ_y, and these two operations in reversed order. Recall that one reads the actions of such operators on an operand on the right, not shown. The key to manipulating such operations is to consider infinitesimal steps: all continuous transformations are effected from a very large number of infinitesimal steps. We illustrate this in detail in the following.

For infinitesimal angles, $\phi_x = \varepsilon_x$, $\phi_y = \varepsilon_y$ (counterclockwise rotations, right-hand rule),

$$R_x(\varepsilon_x) = \exp\{-iL_x\varepsilon_x/\hbar\} \sim I - iL_x\varepsilon_x/\hbar \qquad (A.6)$$

and

$$R_y(\varepsilon_y) = \exp\{-iL_y\varepsilon_y/\hbar\} \sim I - iL_y\varepsilon_y/\hbar, \qquad (A.7)$$

whence

$$R_y(\varepsilon_y)R_x(\varepsilon_x) = \{I - iL_y\varepsilon_y/\hbar\}\{I - iL_x\varepsilon_x/\hbar\}, \qquad (A.8)$$

$$\therefore R_y(\varepsilon_y)R_x(\varepsilon_x) = \{I - iL_y\varepsilon_y/\hbar - iL_x\varepsilon_x/\hbar - L_yL_x\varepsilon_y\varepsilon_x/\hbar^2\}. \qquad (A.9)$$

Similarly,

$$R_x(\varepsilon_x)R_y(\varepsilon_y) = \{I - iL_x\varepsilon_x/\hbar - iL_y\varepsilon_y/\hbar - L_xL_y\varepsilon_x\varepsilon_y/\hbar^2\}, \qquad (A.10)$$

which leads to

$$R_x(\varepsilon_x)R_y(\varepsilon_y) - R_y(\varepsilon_y)R_x(\varepsilon_x) = \cancel{I} - \cancel{iL_x\varepsilon_x/\hbar} - \cancel{iL_y\varepsilon_y/\hbar} - L_xL_y\varepsilon_x\varepsilon_y/\hbar^2$$
$$\cancel{I} + \cancel{iL_y\varepsilon_y/\hbar} + \cancel{iL_x\varepsilon_x/\hbar} + L_yL_x\varepsilon_y\varepsilon_x/\hbar^2, \qquad (A.11)$$

$$\therefore R_x(\varepsilon_x)R_y(\varepsilon_y) - R_y(\varepsilon_y)R_x(\varepsilon_x) = (L_yL_x - L_xL_y)\varepsilon_x\varepsilon_y/\hbar^2$$
$$= -i\hbar L_z\varepsilon_x\varepsilon_y/\hbar^2 \qquad (A.12)$$
$$= R_z(\varepsilon_x\varepsilon_y) - I,$$

i.e., the difference in the order of performance of the infinitesimal rotations is a rotation about the z-axis through the infinitesimal angle $\varepsilon_x\varepsilon_y$. Only the form of this relationship is needed in order to proceed to consideration of rotations in the intrinsic frame. The similar set of operations in the intrinsic frame follows.

Consider, $R_1(\varepsilon_1)R_2(\varepsilon_2) - R_2(\varepsilon_2)R_1(\varepsilon_1)$, and *for the instantaneous orientation* 1-axis with the x-axis, 2-axis with y-axis, 3-axis with z-axis, we can replace $R_2(\varepsilon_2)$ with $R_y(\varepsilon_2)$, for the R_1R_2 (first term), viz.

$$R_1(\varepsilon_1)R_2(\varepsilon_2) = R_1(\varepsilon_1)R_y(\varepsilon_2), \tag{A.13}$$

but then no further replacement is valid because following $R_y(\varepsilon_2)$, the x-axis is no longer collinear with the 1-axis. We designate the newly oriented 1-axis the x'-axis and obtain for the R_1R_2 term

$$R_1(\varepsilon_1)R_2(\varepsilon_2) = R_{x'}(\varepsilon_1)R_y(\varepsilon_2). \tag{A.14}$$

The challenge here is to handle the $R_{x'}(\varepsilon_1)$ term. To do this we use a similarity transformation $R_n^{-1}(\psi_n)R_{x'}(\varepsilon_1)R_n(\psi_n)$. To understand this in practical terms (the concept is equally valid in the space in which we live and in state vector space, i.e. Hilbert space), consider the following task. A book on a tightly packed bookshelf has its title upside-down. Take the book out [translation, T], rotate the title right-side up [rotation, R], return the book to its original position on the shelf [inverse translation, T^{-1}], the result is the execution of $T^{-1} \ R \ T$, a similarity transformation. Above we have defined a similar task, but with a translation replaced by a rotation and its inverse. We need to rotate the x'-axis so that it becomes the x-axis. At the instant after the infinitesimal transformation $R_y(\varepsilon_2)$ was made, the rotation that we seek is $R^{-1} = R_y(\varepsilon_2)$, so $R = R_y(-\varepsilon_2)$. Thus, we obtain

$$R_{x'}(\varepsilon_1) = R_y(\varepsilon_2)R_x(\varepsilon_1)R_y(-\varepsilon_2). \tag{A.15}$$

Whence, for the R_1R_2 term

$$\begin{aligned}
R_1(\varepsilon_1)R_2(\varepsilon_2) &= R_{x'}(\varepsilon_1)R_y(\varepsilon_2) \\
&= R_y(\varepsilon_2)R_x(\varepsilon_1) \, R_y(-\varepsilon_2)R_y(\varepsilon_2) \\
&= R_y(\varepsilon_2)R_x(\varepsilon_1).
\end{aligned} \tag{A.16}$$

Similarly, for the R_2R_1 term, noting that the 1-axis = x-axis rotation occurs first for this term

$$R_2(\varepsilon_2)R_1(\varepsilon_1) = R_x(\varepsilon_1)R_y(\varepsilon_2). \tag{A.17}$$

It then follows by inspection that

$$\begin{aligned}
R_1(\varepsilon_1)R_2(\varepsilon_2) - R_2(\varepsilon_2)R_1(\varepsilon_1) &= R_y(\varepsilon_2)R_x(\varepsilon_1) - R_x(\varepsilon_1)R_y(\varepsilon_2) \\
&= - \{R_x(\varepsilon_1)R_y(\varepsilon_2) - R_y(\varepsilon_2)R_x(\varepsilon_1)\},
\end{aligned} \tag{A.18}$$

i.e. a minus sign is involved for the body-frame commutator brackets, cf equation (A.18) with equation (A.12) and (A.1); thus, equation (A.2) follows.

IOP Publishing

Nuclear Data
A collective motion view
David Jenkins and John L Wood

Appendix B

A generic two-band mixing formalism

Mixing of states with the same spin-parity in two or more bands occurs widely in nuclei. This has been addressed in a specific situation under the title of Mikhailov theory in section 2.4. Here, a general formalism is presented which is designed to keep track of magnitudes of the matrix elements of the electric quadrupole, $E2$ and electric monopole, $E0$ operators and how they contribute to electromagnetic transition strengths.

For the mixing of two configurations of spin J, from bands a and b,

$$| J_1 \rangle = \alpha_J | J^a \rangle + \beta_J | J^b \rangle, \qquad | J_2 \rangle = -\beta_J | J^a \rangle + \alpha_J | J^b \rangle, \tag{B.1}$$

can be written, where the mixing amplitudes obey $\alpha_J^2 + \beta_J^2 = 1$.

For $E2$ properties, intrinsic matrix elements,

$$M_{20}^a = \langle 2^a | T(E2) | 0^a \rangle, \tag{B.2}$$

$$M_{20}^b = \langle 2^b | T(E2) | 0^b \rangle, \tag{B.3}$$

are introduced, and then all other matrix elements are defined by the axially symmetric rotor model using Clebsch–Gordan coefficients, viz.

$$M_{20}^a = \sqrt{B(E2; 0_a^+ \to 2_a^+)}, \tag{B.4}$$

$$M_{J, J-2}^a = \frac{\sqrt{3J(J-1)}}{\sqrt{2(2J-1)}} M_{20}^a, \tag{B.5}$$

$$M_{J, J}^a = -\frac{\sqrt{J(J+1)(2J+1)}}{\sqrt{(2J-1)(2J+3)}} M_{20}^a, \tag{B.6}$$

with similar expressions for band b. The $E2$ matrix elements for the resulting mixed bands are then obtained directly, viz.

doi:10.1088/978-0-7503-5643-5ch9

$$M_{2_1 0_1} = \alpha_0\alpha_2 M_{20}^a + \beta_0\beta_2 M_{20}^b, \qquad (B.7)$$

$$M_{4_1 2_1} = (\alpha_2\alpha_4 M_{20}^a + \beta_2\beta_4 M_{20}^b)(1.604), \qquad (B.8)$$

$$M_{2_2 0_2} = \beta_0\beta_2 M_{20}^a + \alpha_0\alpha_2 M_{20}^b, \qquad (B.9)$$

$$M_{2_2 0_1} = -\alpha_0\beta_2 M_{20}^a + \alpha_2\beta_0 M_{20}^b, \qquad (B.10)$$

$$M_{2_2 2_1} = \alpha_2\beta_2(M_{20}^a - M_{20}^b)(-1.195), \qquad (B.11)$$

$$M_{0_2 2_1} = -\alpha_2\beta_0 M_{20}^a + \alpha_0\beta_2 M_{20}^b. \qquad (B.12)$$

These matrix elements follow from the substitution of equations (B.1) and (B.2) into, e.g.

$$M_{2_1 0_1} = \langle 2_1 | T(E2) | 0_1 \rangle, \qquad (B.13)$$

etc and adopting zero values for the inter-band matrix elements.

For $E0$ properties, the intrinsic matrix elements are introduced

$$\langle r^2 \rangle_a = \langle 0^a | r^2 | 0^a \rangle, \qquad (B.14)$$

$$\langle r^2 \rangle_b = \langle 0^b | r^2 | 0^b \rangle, \qquad (B.15)$$

and zero values for the inter-band matrix elements are adopted.

Note the following:

1. The minus signs in (B.10)–(B.12) result in cancellations (destructive interference);

2. table B.1 reveals a natural hierarchy of inter-band transition strengths which match the cancellations.

The $E0$ properties for the resulting mixed bands are then obtained directly, viz.

$$\rho_J(E0) = \alpha_J\beta_J(\langle r^2 \rangle_a - \langle r^2 \rangle_b) = \alpha_J\beta_J\Delta\langle r^2 \rangle, \qquad (B.16)$$

$$\delta\langle r^2 \rangle_{2_1 0_1} = \left(\beta_2^2 - \beta_0^2\right)\Delta r^2. \qquad (B.17)$$

The strength of $E0$ transitions is expressed as

$$\rho_J^2 10^3 = \alpha_J^2\beta_J^2(\Delta\langle r^2 \rangle)^2 10^3 Z^2/R^4, \qquad (B.18)$$

where $R = 1.2A^{1/3}$ fm, and the factor 10^3 is by convention. The quantity $\delta\langle r^2 \rangle_{2_1 0_1}$ is sometimes called the isomer shift (between the ground and the first excited state).

Table B.1. Details of the mixing for the $K = 0_1$ (ground-state band) and $K = 0_2$ (first excited $K = 0$) band in ^{152}Sm. The parameters are $M_{20}^a = 1.650$ e.b, $M_{20}^b = 2.300$ e.b and the amplitudes $\alpha_0 = 0.8458$, etc given in the box.

Two-band mixing calculation for E2 properties of ^{152}Sm

$Q(2_1^+)$

expt.	-1.683^{18}	b
calc.	-1.690	b

$$B(E2) = \frac{M^2}{2J_i + 1} \times 207.5 \ W.u.$$

Rotor matrix elements

$$M_{J,J-2} = \sqrt{\frac{3J(J-1)}{2(2J-1)}} M_{20}$$

$$M_{JJ} = -\sqrt{\frac{J(J+1)(2J+1)}{(2J-1)(2J+3)}} M_{20}$$

$$M_{20} = \sqrt{B(E2; 0_1^+ \to 2_1^+)}$$

$$M_{20}^a = 1.650 \quad e \cdot b$$

$$M_{20}^b = 2.300 \quad e \cdot b$$

	calc.	exp.
$2_1 \to 0_1$	141.7	145.0^{16}
$4_1 \to 2_1$	210.8	209.5^{22}
$6_1 \to 4_1$	245.7	240^4
$8_1 \to 6_1$	271.6	293^4
$2_2 \to 0_2$	182.5	170^{12}
$4_2 \to 2_2$	249.4	250^{40}
$0_2 \to 2_1$	33.9	33.3^{12}
$2_2 \to 4_1$	24.5	18.0^{12}
$4_2 \to 6_1$	23.1	17^3
$2_2 \to 2_1$	5.56	5.7^4
$4_2 \to 4_1$	5.52	5.0^{+10}_{-7}
$2_2 \to 0_1$	1.61	0.94^6
$4_2 \to 2_1$	1.32	0.74^{12}

e.g.,

$M_{2_1 0_1}$	$=$	$\alpha_0 \alpha_2 M_{20}^a + \beta_0 \beta_2 M_{20}^b$	$=$	$1.848 \ e \cdot b$
$M_{4_1 2_1}$	$=$	$(\alpha_2 \alpha_4 M_{20}^a + \beta_2 \beta_4 M_{20}^b)(1.604)$	$=$	3.024
$M_{2_2 0_2}$	$=$	$\beta_0 \beta_2 M_{20}^a + \alpha_0 \alpha_2 M_{20}^b$	$=$	2.097
$M_{2_2 0_1}$	$=$	$-\alpha_0 \beta_2 M_{20}^a + \alpha_2 \beta_0 M_{20}^b$	$=$	0.197
$M_{2_2 2_1}$	$=$	$\alpha_2 \beta_2 (M_{20}^b - M_{20}^a)(-1.195)$	$=$	-0.366
$M_{0_2 2_1}$	$=$	$-\alpha_2 \beta_0 M_{20}^a + \alpha_0 \beta_2 M_{20}^b$	$=$	0.404

J	0	2	4	6	8
α_J	0.8458	0.8167	0.7656	0.7071	0.6536
β_J	0.5334	0.5770	0.6433	0.7071	0.7569

B.1 Exercises

B-1. Derive equations for $M_{4_1 2_2}$, $M_{4_2 2_2}$, $M_{4_2 2_1}$, $M_{4_2 4_1}$, $M_{4_1 4_1}$, $M_{4_2 4_2}$.

B-2. Derive equations for $M_{6_1 4_1}$, $M_{6_1 6_1}$.

B-3. Derive equation (B.18).

B-3. Using data for ^{152}Sm in ENSDF, what is the Grodzins product? (Note that the product is for states which are mixed configurations with different deformations.)

B-4. Complete all of the computational steps for the calculated (calc.) values in table B.1, using the input values given for mixing amplitudes and model parameters.

B-5. What modifications would be needed to the above formalism to handle the mixing of two $K = 2$ bands?

B-6. What modifications would be needed to the above formalism to handle the mixing of a $K = 0$ band with a $K = 2$ band?

IOP Publishing

Nuclear Data
A collective motion view
David Jenkins and John L Wood

Appendix C

$E2$ matrix elements for selected even–even nuclei and selected transitions

See tables C.1, C.2, C.3, C.4, C.5 and C.6 for selected even–even nuclei and selected transitions.

Table C.1. Compilation of $E2$ matrix elements in units of e.b for selected transitions in selected even–even zinc ($Z = 30$) and germanium ($Z = 32$) nuclei. The values for $0_1 \rightarrow 2_1$, shown in blue, are derived from $B(E2; 0_1 \rightarrow 2_1)$ values in $e^2 \cdot b^2$ given in [1]. Other data are taken from: ^{66}Zn [2], ^{68}Zn [3], ^{70}Ge [4], ^{72}Ge [5], ^{74}Ge [6], and ^{76}Ge [7].

	^{66}Zn	^{68}Zn	^{70}Ge	^{72}Ge	^{74}Ge	^{76}Ge
$0_1 \rightarrow 2_1$	$+0.370^4$	$+0.346^3$	$+0.423^4$	$+0.457^3$	$+0.553^{14}$	$+0.523^3$
$0_1 \rightarrow 2_2$	$+0.0048^7_5$	$+0.069^3$	-0.0434^{13}	$+0.030^1$	$+0.058^{10}$	$+0.089^3$
$2_1 \rightarrow 2_2$	$+0.57^{10}$	-0.39^4	$+0.42^7$	$+0.65^1_2$	$+0.50^4$	$+0.535^3_7$
$2_1 \rightarrow 2_1$	$+0.32^{10}$	$+0.12^4$	$+0.05^4$	-0.16^7_2	-0.25^3	-0.24^2
$2_2 \rightarrow 2_2$		$+0.12^8$	-0.09^5	$+0.179^3_6$	$+0.34^8$	$+0.26^2_5$
$2_1 \rightarrow 4_1$	$+0.500^{10}$	$+0.441^7$	$+0.54^{10}$	$+0.90^2$	$+0.850^{25}$	$+0.795^5$
$4_1 \rightarrow 6_1$	$+1.39^6$			$+1.11^4_5$		$+1.11^3_2$
$4_1 \rightarrow 4_1$			$+0.29^7$	-0.14^9_4		-0.26^1_7
$6_1 \rightarrow 6_1$				-0.20^8_{25}		-0.23^9_4
$2_1 \rightarrow 4_2$				$+0.035^6$		-0.22^5_3
$2_2 \rightarrow 4_1$		$+0.31^5$	-0.52^{12}	-0.06^3_4	$+0.05^{25}$	$+0.09^2$
$2_2 \rightarrow 4_2$				$+0.58^5_1$		$+0.472^6$
$4_1 \rightarrow 4_2$				$+0.43^{10}$		$+0.61^1$

doi:10.1088/978-0-7503-5643-5ch10

Table C.2. Compilation of $E2$ matrix elements in units of e.b for selected transitions in selected even–even selenium ($Z = 34$) nuclei. The values for $0_1 \rightarrow 2_1$, shown in blue, are derived from $B(E2; 0_1 \rightarrow 2_1)$ values in e^2· b^2 given in [1]. Other data are taken from: ^{76}Se [8], ^{78}Se [9], and ^{80,82}Se [10].

	^{76}Se	^{78}Se	^{80}Se	^{82}Se
$0_1 \rightarrow 2_1$	$+0.657^{11}_{5}$	$+0.586^{10}$	$+0.502^{8}$	$+0.428^{12}$
$0_1 \rightarrow 2_2$	$+0.110^{2}$	$+0.08^{1}$	$+0.106^{6}$	$+0.120^{6}$
$2_1 \rightarrow 2_2$	$+0.640^{11}$	$+0.45^{4}$	$+0.38^{2}$	$+0.19^{2}$
$2_1 \rightarrow 2_1$	-0.46^{5}	-0.27^{9}	-0.26^{4}_{3}	-0.30^{4}_{3}
$2_2 \rightarrow 2_2$	$+0.25^{6}$	$+0.23^{12}$	$+0.53^{3}$	$+0.45^{4}_{5}$
$2_1 \rightarrow 4_1$	$+1.108^{12}_{11}$	$+0.81^{6}$	$+0.82^{4}$	$+0.63^{3}$
$4_1 \rightarrow 6_1$	$+1.39^{6}$		$+1.14^{29}_{15}$	
$4_1 \rightarrow 4_1$	-0.39^{6}_{5}	-0.90^{20}	-0.85^{11}_{6}	-0.76^{7}_{8}
$6_1 \rightarrow 6_1$				
$2_1 \rightarrow 4_2$	$+0.039^{35}_{7}$		$\pm0.01^{13}_{6}$	$+0.09^{1}$
$2_2 \rightarrow 4_1$	$+0.05^{4}_{3}$	$\pm0.09^{4}_{5}$	$+0.08^{4}_{13}$	
$2_2 \rightarrow 4_2$	$+0.77^{4}$		$+0.67^{8}_{18}$	$+0.71^{3}_{9}$
$4_1 \rightarrow 4_2$	$+0.73^{5}_{4}$			$+0.28^{5}_{4}$

Table C.3. Compilation of $E2$ matrix elements in units of e.b for selected transitions in selected even–even Krypton ($Z = 36$) nuclei. The values for $0_1 \rightarrow 2_1$, shown in blue, are derived from $B(E2; 0_1 \rightarrow 2_1)$ values in e^2· b^2 given in [1]. Other data are taken from: ^{74,76}Kr [11], ^{78}Kr [12], ^{80}Kr [13], ^{82}Kr [14], and ^{84}Kr [15].

	^{74}Kr	^{76}Kr	^{78}Kr	^{80}Kr	^{82}Kr	^{84}Kr
$0_1 \rightarrow 2_1$	$+0.792^{20}$	$+0.871^{15}$	$+0.796^{10}$	$+0.617^{10}$	$+0.474^{7}$	$+0.356^{16}$
$0_1 \rightarrow 2_2$	-0.199^{18}_{11}	$+0.183^{8}_{6}$	$+0.157^{17}_{4}$	$+0.078^{15}$	-0.035^{11}_{8}	0.17^{2}
$2_1 \rightarrow 2_2$	$+0.49^{4}$	-0.09^{4}	$+0.26^{6}_{5}$	$+0.73^{14}$	-0.28^{91}_{6}	0.35^{14}
$2_1 \rightarrow 2_1$	-0.7^{3}	-0.9^{3}	-0.80^{4}	-0.43^{7}_{3}		
$2_2 \rightarrow 2_2$	$+0.3^{3}_{2}$	-1.0^{5}	$+0.58^{4}_{8}$	$+0.4^{17}$		
$2_1 \rightarrow 4_1$	$+1.60^{3}$	$+1.49^{1}$	$+1.27^{3}_{3}$		$+0.78^{13}_{16}$	$+0.69^{9}$
$4_1 \rightarrow 6_1$	$+1.98^{10}_{9}$	$+1.90^{11}_{3}$	$+1.61^{6}_{8}$	$+1.68^{12}$	$+0.74^{17}_{23}$	
$4_1 \rightarrow 4_1$	-1.0^{6}_{2}	-2.3^{4}	-0.73^{15}_{14}	-0.77^{22}		
$6_1 \rightarrow 6_1$	-1.8^{7}_{5}	-2.9^{4}				
$2_1 \rightarrow 4_2$		$+0.09^{19}_{1}$	$+0.073^{2}_{5}$		$+0.15^{2}$	
$2_2 \rightarrow 4_1$	$+0.47^{3}_{2}$	-0.62^{4}_{5}	$+0.32^{5}_{4}$			
$2_2 \rightarrow 4_2$	$+0.55^{16}_{8}$		$+0.91^{6}_{4}$		$+0.41^{8}$	
$4_1 \rightarrow 4_2$		$+0.43^{3}$	-0.60^{2}_{3}		$+0.87^{11}$	

Table C.4. Compilation of $E2$ matrix elements in units of e.b for selected transitions in selected even–even ruthenium ($Z = 44$), palladium ($Z = 46$) and cadmium ($Z = 48$) nuclei. The values for $0_1 \rightarrow 2_1$, shown in blue, are derived from $B(E2; 0_1 \rightarrow 2_1)$ values in $e^2 \cdot b^2$ given in [1]. Other data are taken from: ^{104}Ru [16], ^{106,108}Pd [17], ^{110}Pd [18], and ^{114}Cd [19].

	^{104}Ru	^{106}Pd	^{108}Pd	^{110}Pd	^{114}Cd
$0_1 \rightarrow 2_1$	$+0.909^9$	$+0.812^{20}$	$+0.874^{11}$	$+0.930^{12}$	$+0.732^{17}$
$0_1 \rightarrow 2_2$	-0.156^2	-0.114^6	-0.98^5	-0.096^2_3	$+0.091^3$
$2_1 \rightarrow 2_2$	-0.75^4	-0.76^4	-0.88^4	-0.863^{11}_{16}	$+0.684^{21}$
$2_1 \rightarrow 2_1$	-0.71^{11}	-0.72^4_7	-0.81^4_9	-0.87^{17}_{15}	-0.36^1_3
$2_2 \rightarrow 2_2$		$+0.52^6_5$	$+0.73^9_7$	$+0.70^9_{32}$	$+0.92^4_5$
$2_1 \rightarrow 4_1$	$+1.43^4$	$+1.38^7$	$+1.42^7$	$+1.579^4_{37}$	$+1.35^4$
$4_1 \rightarrow 6_1$	$+2.04^8$	$+1.86^{10}_{14}$	$+2.06^{11}$	$+2.08^8_3$	$+2.3^3$
$4_1 \rightarrow 4_1$	-0.79^{15}	-1.02^7_{11}	-0.78^{11}_{10}	-1.6^4_2	-0.95^4_{11}
$6_1 \rightarrow 6_1$	-0.7^3_2	-1.41^{23}_{13}	-0.76^{18}	-1.4^2_4	-3.5^9
$2_1 \rightarrow 4_2$	-0.107^8	-0.014^5_4		-0.066^{15}_{12}	$+0.11^1$
$2_2 \rightarrow 4_1$		$+0.143^{30}$	$+0.18^9_{13}$	$+0.51^{11}_{32}$	-0.35^7
$2_2 \rightarrow 4_2$	$+1.12^5$	-0.30^8_5	$+1.23^7_6$	$+0.97^4_3$	$+0.97^{17}_3$
$4_1 \rightarrow 4_2$	-0.88^5	$+0.79^4$	-0.91^7_8	-0.94^5_4	$+0.61^8_4$

Table C.5. Compilation of $E2$ matrix elements in units of e.b for selected transitions in selected even–even tellurium ($Z = 52$), xenon ($Z = 54$) and neodymium ($Z = 60$) nuclei. The values for $0_1 \rightarrow 2_1$, shown in blue, are derived from $B(E2; 0_1 \rightarrow 2_1)$ values in $e^2 \cdot b^2$ given in [1]. Other data are taken from: ^{122}Te [18], ^{126,128}Xe [20], and ^{148}Nd [21].

	^{122}Te	^{126}Xe	^{128}Xe	^{148}Nd
$0_1 \rightarrow 2_1$	$+0.806^{19}$	$+0.91^3$	$+0.889^{21}$	$+1.157^{13}$
$0_1 \rightarrow 2_2$	$+0.110^2$	$+0.119^9$	0.105^8	$+0.123^5_4$
$2_1 \rightarrow 2_2$	$+0.640^{11}$	$+1.00^4$	$+0.92^4$	-0.65^2
$2_1 \rightarrow 2_1$	-0.46^5	-1.0^2	-0.58^{12}_{15}	-1.85^4_5
$2_2 \rightarrow 2_2$	$+0.25^6$	$+0.14^9$	$+0.01^9_{10}$	-1.15^{18}_{12}
$2_1 \rightarrow 4_1$	$+1.08^{12}_{11}$	$+1.48^4$	$+1.38^4$	$+2.00^4$
$4_1 \rightarrow 6_1$	$+1.39^6$	$+2.07^9$	$+1.95^{12}$	$+2.62^7$
$4_1 \rightarrow 4_1$	-0.39^6_5	-0.78^{16}	-1.38^{13}	-1.40^{17}
$6_1 \rightarrow 6_1$				-1.72^{19}_{20}
$2_1 \rightarrow 4_2$	$+0.39^{35}_7$			$+0.072^5$
$2_2 \rightarrow 4_1$	$+0.05^4_3$			$+1.12^3_5$
$2_2 \rightarrow 4_2$	$+0.77^4$	$+0.97^6$		$+2.06^8_7$
$4_1 \rightarrow 4_2$	$+0.73^5_4$			-0.338^{20}_{15}

Table C.6. Compilation of $E2$ matrix elements in units of e.b for selected transitions in selected even–even osmium ($Z = 76$) and platinum ($Z = 78$) nuclei. The values for $0_1 \to 2_1$, shown in blue, are derived from $B(E2; 0_1 \to 2_1)$ values in $e^2 \cdot b^2$ given in [1]. Other data are taken from: ^{186,188,190,192}Os [22], ^{194}Pt [22], and ^{196}Pt [23].

	^{186}Os	^{188}Os	^{190}Os	^{192}Os	^{194}Pt	^{196}Pt
$0_1 \to 2_1$	$+1.750^{21}$	$+1.581^{11}$	$+1.53^3$	$+1.43^4$	$+1.277^{27}$	$+1.184^{29}$
$0_1 \to 2_2$	$+0.545^{13}_{7}$	$+0.483^2_9$	$+0.444^9_7$	$+0.430^8_4$	$+0.0888^{12}$	0.000
$2_1 \to 2_2$	$+0.897^{67}_{14}$	$+0.865^{11}$	$+1.065^{20}_{37}$	$+1.230^{34}_{16}$	$+1.517^{11}_{18}$	$+1.36^1$
$2_1 \to 2_1$	-1.75^{22}_{13}	-1.73^{19}_{5}	-1.25^{22}_{13}	-1.21^6_{17}	$+0.54^8_6$	$+0.82^{10}$
$2_2 \to 2_2$	$+2.12^6_{22}$	$+2.10^9_6$	$+1.53^6_{31}$	$+0.99^6_9$	-0.40^{12}_{5}	-0.52^{20}
$2_1 \to 4_1$	$+2.76^6_7$	$+2.642^{25}_{20}$	$+2.37^3$	$+2.12^3$	$+1.935^{21}_{13}$	$+1.91^3$
$4_1 \to 6_1$	$+3.89^8_5$	$+3.31^4$	$+2.97^6_4$	$+2.93^7_4$	$+2.90^{10}_{4}$	$+2.42^7$
$4_1 \to 4_1$	-2.02^{39}_{18}	-2.00^9_{20}	-1.28^{27}_{19}	-0.73^{26}_{6}	$+1.00^{12}_{14}$	$+1.35^{16}$
$6_1 \to 6_1$	-1.67^{29}_{21}	-1.60^{18}_{33}	-0.91^{24}_{15}	-1.16^{11}_{26}	$+0.28^{12}_{27}$	-0.3^4
$2_1 \to 4_2$	$+0.419^{27}_{15}$	$+0.283^8_7$	$+0.203^7$	$+0.130^5_8$	$+0.220^9$	$+0.11^7$
$2_2 \to 4_1$		$+0.38^5_6$	$+0.19^{12}_{9}$	$+0.35^{16}_{4}$	$+0.25^4_6$	
$2_2 \to 4_2$	$+1.97^9_7$	$+1.78^5_3$	$+1.87^4$	$+1.637^{24}_{33}$	$+1.78^5_3$	$+1.28^6$
$4_1 \to 4_2$	$+1.22^6$	$+1.10^3$	$+1.44^4$	$+1.35^8_4$	$+1.51^6_5$	$+0.87^7$

C.1 Exercises

C-1. Test the triangular relationships depicted in figures 2.5(a) and (b) for deduced Q_0 values. Example: for ^{188}Os, $(0.865^2 + 1.73^2) \times 56\pi/25 = 5.13$ e.b cf $(1.581^2 + 0.483^2) \times 16\pi/5 = 5.24$ e.b. (Note that the signs on the matrix elements are ignored because they are being combined in quadrature.)

C-2. Check the agreement of the Q_0 values deduced from the two triangle relationships within the experimental uncertainties given, e.g. $0.865^{11} = 0.865 \pm 0.011$.

References

[1] Pritychenko B, Birch M, Singh B and Horoi M 2016 tables of $E2$ transition probabilities from the first 2^+ states in even-even nuclei *At. Data Nucl. Data Tables* **107** 1–139

[2] Rocchini M *et al* 2021 Onset of triaxial deformation in ^{66}Zn and properties of its first excited 0^+ state studied by means of Coulomb excitation *Phys. Rev. C* **103** 014311

[3] Koizumi M *et al* 2004 Multiple Coulomb excitation experiment of ^{68}Zn *Nucl. Phys.* A **730** 46

[4] Sugawara M 2003 Multiple Coulomb excitation of a ^{70}Ge beam and the interpretation of the 0_2^+ state as a deformed intruder *Eur. Phys. J.* A **16** 409–14

[5] Ayangeakaa A D *et al* 2016 Shape coexistence and the role of axial symmetry in ^{72}Ge *Phys. Lett.* B **754** 254–9

[6] Toh Y *et al* 2000 Coulomb excitation of ^{74}Ge beam *Eur. Phys. J.* A **9** 353–6

[7] Ayangeakaa A D *et al* 2023 Triaxiality and the nature of low-energy excitations in ^{76}Ge *Phys. Rev.* C **107** 044314

[8] Henderson J *et al* 2019 Triaxiality in selenium-76 *Phys. Rev.* C **99** 054313

[9] Hayakawa T *et al* 2003 Projectile Coulomb excitation of ^{78}Se *Phys. Rev.* C **67** 064310

[10] Kavka A E *et al* 1995 Coulomb excitation of ^{76,80,82}Se *Nucl. Phys.* A **593** 177–211

[11] Clément E *et al* 2007 Shape coexistence in neutron-deficient krypton isotopes *Phys. Rev.* C **75** 054313

[12] Becker F *et al* 2006 Coulomb excitation of ^{78}Kr *Nucl. Phys.* A **770** 107–25

[13] Gillespie S A *et al* 2021 Coulomb excitation of ^{80,82}Kr and a change in structure approaching $N = Z = 40$ *Phys. Rev.* C **104** 044313

[14] Brüssermann S, Lieb K P, Sona P, Emling H, Grosse E and Stachel J 1985 Multiple Coulomb excitation of the transitional nucleus ^{82}Kr *Phys. Rev.* C **32** 1521

[15] Osa A *et al* 2002 First measurement of the quadrupole moment in the 2_1^+ state of ^{84}Kr *Phys. Lett.* B **546** 48–54

[16] Srebrny J *et al* 2006 Experimental and theoretical investigations of quadrupole collective degrees of freedom in ^{104}Ru *Nucl. Phys.* A **766** 25–51

[17] Svensson L E *et al* 1995 Multiphonon vibrational states in ^{106,108}Pd *Nucl. Phys.* A **584** 547–72

[18] Svensson L E 1989 Coulomb excitation of vibrational nuclei *PhD thesis* University of Uppsala

[19] Fahlander C *et al* 1988 Quadrupole collective properties of ^{114}Cd *Nucl. Phys.* A **485** 327–59

[20] Kisyov S *et al* 2022 Structure of ^{126,128}Xe studied in coulomb excitation measurements *Phys. Rev.* C **106** 034311

[21] Ibbotson R W *et al* 1997 Quadrupole and octupole collectivity in ^{148}Nd *Nucl. Phys.* A **619** 213–40

[22] Wu C Y *et al* 1996 Quadrupole collectivity and shapes of Os-Pt nuclei *Nucl. Phys.* A **607** 178–234

[23] Lim C S, Spear R H, Fewell M P and Gyapong G J 1992 Measurements of static electric quadrupole moments of the 2_1^+, 2_2^+, 4_1^+ and 6_1^+ states of ^{196}Pt *Nucl. Phys.* A **548** 308

Printed in the USA
CPSIA information can be obtained
at www.ICGtesting.com
JSHW061315120124
55268JS00004B/40

9 780750 356411